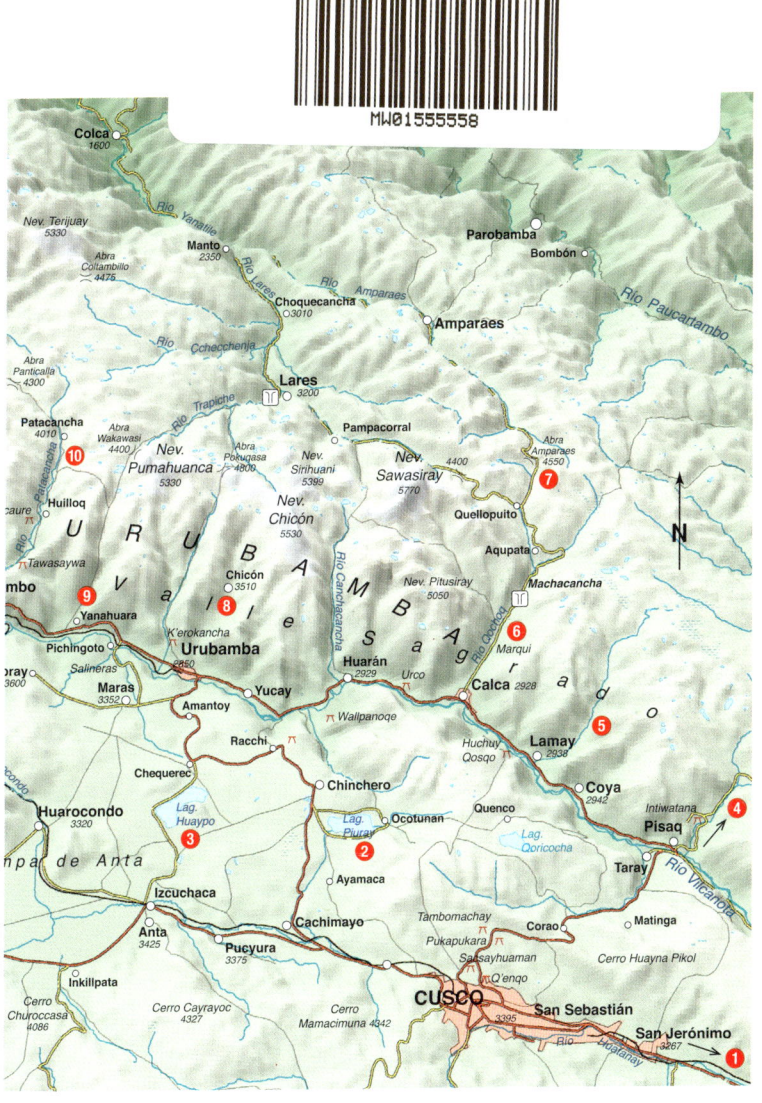

Leyenda / Legend

— Vía asfaltada / Paved Road
— Vía carrozable / Unpaved Road
— Camino / Trail
— Ferrovía / Railway
○ ⋔ Centro poblado, Ruina / Populated center, Ruin

This map is a portion of the "Camino Inka/Inka Trail" map published by Lima 2000, reprinted with permission. See back endpapers for ITMB map of the Inca Trail.

FIELD GUIDE TO THE BIRDS OF
MACHU PICCHU AND THE CUSCO REGION

FIELD GUIDE TO THE BIRDS OF MACHU PICCHU AND THE CUSCO REGION

Includes a Bird Finding Guide to the Area

Barry Walker

First Edition: June 2015

© Lynx Edicions – Montseny, 8, 08193 Bellaterra, Barcelona, www.lynxeds.com
© Buteo Books – 2731 Arrington Road, Arrington, VA 22922, USA, www.buteobooks.com
© Texts and photographs: Barry Walker
© Illustrations: Lynx Promocins, S. L.
© Cover illustration: Jon Fjeldså
© Illustrations of the following species: Francesc Jutglar (VEGAP, Barcelona 2015). *Calidris alba* non breeding, *Pluvialis squatarola* non breeding, *Calidris melanotos* non breeding, *Chlidonias niger surinamensis* non breeding, *Pandion haliaetus carolinensis* flying, *Circus cinereus* juvenile flying, *Falco peregrinus cassini* flying, *Falco peregrinus cassini* juvenile flying, *Calidris fuscicollis* non breeding i *Falco peregrinus tundrius* flying.

Design and layout: Lynx Edicions

Printed by: Ingoprint, S.A.
Legal Deposit: B-7084-2015
ISBN: 978-84-96553-97-2

All rights reserved. No form of reproduction, distribution, public communication or transformation of this work may be carried out without the authorization of its copyrights holders, except that foreseen by the law. Those needing to photocopy or electronically scan any part of this work should contact Lynx Edicions.

CONTENTS

Preface 7

Acknowledgements 9

An Introduction to the Machu Picchu area (*by Jon Fjeldså*) 11

Birding Machu Picchu and the Cusco Region (*by Huw Lloyd*) 15

Habitats and micro-habitats of the Machu Picchu and Cusco area 23

Bird species of special interest and concern in the area 25

Species accounts 31

A bird finding guide to the Cusco and Machu Picchu area 201

An annotated checklist of the birds of Machu Picchu and Cusco 215

References 235

Index 237

PREFACE

The publication of a *Field Guide to the Birds of Machu Picchu and the Cusco Region* is the result of a remarkable three-continent collaboration between Barry Walker in Peru, Buteo Books in Virginia (USA) and Lynx Edicions in Barcelona. Previous versions of Barry Walker's field guide were published in Peru, but efforts to find electronic files of the text and illustrations met with no success. Thus, we were faced with the challenge of preparing an entirely new book with updated text and new illustrations. Lynx Edicions offered the use of illustrations painted for the monumental *Handbook of the Birds of the World*. Additional illustrations have been painted by Matin Elliott of birds in flight and subspecies present in the Cusco Region. At Buteo Books, Allen Hale undertook the task of editing the text so that species accounts would fit on pages facing the illustrations while maintaining the description of the bird's status and distribution, behavior, habitat and song as well as key field marks. The final layout and assembly were accomplished by the staff at Lynx Edicions.

Throughout the process, our goal was to produce a field guide with all the essentials to identify the bird in the field, yet compact and light enough to fit in a pocket or carry on the Inca Trail. While the main body of the book is devoted to the species descriptions and their illustrations, the "Introduction" by Jon Fjeldså and "Birding Machu Picchu and the Cusco Region" by Huw Lloyd provide additional information on this biologically rich region. Also, Barry Walker has written a detailed guide to the best birding sites in the region that will be of great interest to the visiting birdwatcher. A complete "Annotated Checklist" of the birds of the region and a select bibliography are included. Portions of maps from ITMB Publishing/International Travel Maps and Lima 2000 have been reprinted on the endpapers, further enhancing the utility of the guide.

A *Field Guide to the Birds of Machu Picchu and the Cusco Region* joins the text of an experienced field ornithologist and guide with illustrations by a number of the best bird artists in the world. Visitors from all over the world who travel to Cusco and the Sacred Valley of the Incas with the ultimate goal of seeing the extraordinary ruins of Machu Picchu will now have a field guide to the beautiful and varied bird life inhabiting the region.

ACKNOWLEDGEMENTS

The author would like to thank the following people who provided invaluable help in the creation and publication of this book: His Excellency Mikko Pyhala, former Finnish Ambassador to Peru, for inspiring earlier editions of this field guide, and to Manuel A. Plenge for his continued support in sourcing data. Thanks to Allen Hale at Buteo Books for suggesting a third edition and to all at Lynx Editions, especially Josep del Hoyo and Susanna Silva, for support with the artwork. Previously unpublished, new distributional data was provided by the late Jim Clements, Gunnar Engblom, Tor Egil Hogsas, Erik Mollgard, Julio Ochoa Estrada, Lucila Pautrat, Omar Rocha, John P. O'Neill, Paul Salaman, Thomas S. Schulenberg, Carl Christian Tofte and Willem-Pier Vellinga, Jose Luis Holguin, Braulio Puma, Dennis Osorio Malaga, Juan Carlos Ancco Callapiña, Jose Lavilla, John Rowlett, Byron Palacios, Andy Whittaker, Guy Kirwan, Saturnino Llactahuaman, Pepe Rojas, Ryan Tyrill, Hendrik Torres, Juan Cardenas, Fernando Angulo, Carmen Soto, David Ricalde, Steve Sánchez Calle, Richard Webster, Miguel Arcangel Lezama Ninancuro, Alfredo Begazo, Ramiro Yabar, Berioska Quispe, Eustace Barnes, Colin Bushell, Dan Lane, Steve Hilty, David Wolf, Kevin Zimmer, John Arvin, Wim ten Have, David Geale, Gregorio Ferro, Fabrice Schmitt, Gustavo Bautista, Maura Jurado and special thanks to José Luis Venero for his advice on Huacarpay Lake records. Special thanks also Jose Koechlin for his hospitality at the Inkaterra hotels at Machu Picchu. Very special thanks to Huw Lloyd for agreeing to provide the introductory notes to the book.

Barry Walker

As a bookseller, Buteo Books found it difficult to obtain sufficient copies of previous editions of Barry Walker's *Field Guide to the Birds of Machu Picchu and the Cusco Region* to meet the demand from travelers headed to Peru. So, on my first visit to Peru in 2000 (thanks to my friend and book collector Dominique Alessandri), I met Barry Walker at his office in Cusco and thus began the process to publish the book you see before you today.

Having no electronic version of the text and images from the edition published in 2005, we faced the formidable task of re-typing the entire text, an accomplishment arranged by book designer John Wattai at the Pemcor Printing Company in Lancaster, Pennsylvania. John, who incidentally collects field guides to birds from all over the world, provided many helpful suggestions as we sought a solution for the artwork.

A significant breakthrough came about from a suggestion made by Jonathan Alderfer at a book signing in Charlottesville, Virginia, when I lamented on the difficulties we were encountering with the artwork for the field guide. He suggested we contact Lynx Edicions to see about using images from the *Handbook of the Birds of the World*. That led to what has been a productive and rewarding collaboration between publishers/booksellers Lynx Edicions and Buteo Books.

John Rowlett reviewed the annotated checklist and made valuable suggestions on nomenclature and illustrations. Guy Tudor provided a sympathetic ear on the trials and tribulations of publishing a field guide. Jon Fjeldså willingly and promptly produced an original painting for the cover. Elisabeth (Jamie) Hale, my successor at Buteo Books, provided technical and design skills to meet the challenges encountered as the various parts of the book were assembled. Susanna Silva and the staff at Lynx Edicions were endlessly helpful and patient during the editing process.

Allen Hale
(May 2015)

AN INTRODUCTION TO THE MACHU PICCHU AREA

by Jon Fjeldså

The eastern slope of the Andes toward the Amazon Basin is the biologically richest area on earth. The lowland forests near the foot of the Andes rank highest in terms of the number of species that can be found in one place (point diversity), but the slopes of the Andes are richer in terms of species turnover over a more extensive area (landscape diversity). More than 1000 bird species can be found along a 200 km transect from the western edge of the Amazonian floodplain to the eastern Andean ridge top, a figure comparable to the number of species known to exist throughout the entire Amazonian floodplain of five million square kilometers! As we ascend the Andes, the most important environmental variable affecting vegetation and bird life is the drop in temperature of 0.6°C every 100 m (this change can be locally moderated in places like Machu Picchu, on the transition to warm rain shadow valleys). As humid air rises up from the Amazonian lowlands and is cooled, the terrain becomes wrapped in clouds for much of the time. Although the period from May to September is relatively dry, the interior of the cloud forest above 2500 m remains cool and moist.

In contrast, the barren areas well above 4000 m are subject to extreme variations, from intense solar heat during the day to biting frost at night.

As we ascend from the lowlands, the forest structure changes quite markedly at 2000-2500 m, from relatively tall (20-30 m) pre-montane forest to low montane forest. In the lower one, the tree trunks are often straight and smooth, and a characteristic feature is the presence of fast-growing pioneer trees with large silvery leaves: *Cecropias*. The montane cloud forest, on the other hand, is low and often impenetrably dense, with gnarled trees and abundant epiphytes, mosses, ferns, orchids and large fountains of bromeliad leaves. A variety of colorful tanagers are seen, often in mixed-species flocks, while the more dull-colored birds of the forest understory and of the bamboo thickets most often reveal themselves by their voices.

The upper cloud forest is also known as elfin forest. It is draped with lichens and the foliage often consists of dense, small, leathery leaves. Similar sclerophyllous forests can also be found in places with persistent mist formation on adjacent rain-shadow slopes.

The foliage of the trees condenses moisture in the form of the fine droplets that comprise the mist. In this way these forests serve an immensely important role as a source of water in the montane basins during the dry season.

Above 3800 m the forest is composed mostly of one kind of tree, *Polylepis*, a species characterized by its small leaves and finely laminated red bark. The birds of these forests are modest in color, but count among their number some of the rarest of Andean species.

Unfortunately, most of the terrain at these elevations is treeless due to frequent burning, overgrazing and the lack of forest regeneration. The humid highlands, known as 'paramo', have a spongy vegetation of mosses, tall grasses and low shrubs, while the drier parts of the highland consist of monotonous bunchgrass vegetation ('puna'). Above 4400 m the vegetation becomes very sparse, mostly prostrate rosettes and cushion plants adapted to tolerate the intense radiation from the sun, frost and snow. Glaciers are limited today to areas above 5000 m.

Machu Picchu and the surrounding mountains are renowned for their numerous endemic and near endemic species of birds, plants and other organisms (by endemic, we mean species that exist nowhere else).

This may be due to the special landscape features of the area. While most parts of the east Andean slope incline more or less directly towards the lowlands, the area between Paucartambo and the Ene/Apurímac River forms a large fan of projecting mountain ridges separated by deep valleys. Thus, Machu Picchu overlooks the valley between the Cordillera Vilcanota (which boasts several peaks of around 5700 m) and the Cordillera Vilcabamba (with peaks reaching to well above 6000 m). An analysis of ten year of data from metrological satellites reveals that the mountain ridges offer efficient protection against the impact of cold south polar winds ('friajes') (Fjeldså *et al.* 1999). This suggests that endemic species represent relict populations which could survive periods of climatic instability only in places which were well protected against extreme weather. By retaining relict populations, these more sheltered places played a key role in the evolution of Andean avifauna. Local aggregates of endemic birds are often immediately adjacent to densely populated areas and centers of past advanced civilizations. Special local conditions protecting relict species through periods of environmental stress may also have facilitated the increase of human settlement in the Andes.

Crop predictability may have been a major prerequisite for the transition from a life as hunter-gatherers to sedentary farming systems, and for subsequent technological advances in agriculture. The pattern of human settlement is also affected by the presence of fertile soils in the transition zone (between humid and rain shadow areas), where there is a balance between soil leaching and evaporation. Machu Picchu marks the gateway between the humid lower Urubamba basin and the favorable climate of the Vilcanota valley, which was the center of the Inca culture's great empire. It is probably no coincidence that some of the rarest Andean birds live in the small patches of forest that surround the more inaccessible ancient ruins and terraces of the Vilcanota valley.

So far, these relationships have been overlooked by conservation biologists. In the future, one of the main challenges for conservationists will be how to find ways to conserve biodiversity in areas adjacent to dense rural populations.

THREATS TO THE ANDEAN BIRDS

Being steep and inaccessible, much of the Andean cloud forest zone is virtually uninhabited. However, this does not mean that the forest is undisturbed. In fact, the slopes are often scarred by small or large landslides, which give rise to a natural succession of bushy growth and bamboo thickets. High precipitation and frequent habitat disturbance of this kind help to maintain high biological diversity. The cloud forest is not suited to agriculture. It is wet, cool, and steep, and its shallow soils are rapidly washed away when the earth is exposed by deforestation. Fields can usually only be cultivated for a couple of years, before being left fallow for several years in order to regenerate. A modest human habitat clearing is not very different in effect from the natural disturbance of landslides. However, as new roads facilitate colonization, large areas are transformed into pasture and dense low shrubbery known as 'purma,' as seen throughout the Vilcabamba valley.

The most drastic habitat transformation in the Andes seems to have taken place thousands of years ago, in the highlands and montane basins. Studies of plant remains in lake sediments near the mountain pass above Ollantaytambo, for example, suggest that the area was totally deforested and severely degraded some 1000-4000 years ago (Chepstow-Lusty *et al.* 1998). Subsequently, agroforestry systems were apparently introduced by the Incas, and the area was brought into a sustainable footing again until their land management systems were destroyed by the Spanish conquest.

Today, most of the Cusco area is severely degraded. The highlands are severely affected by the frequent burning used to maintain pasture for sheep and cattle, and some areas seem to be burned for no obvious purpose. Currently, tree line is located several hundred meters below its natural level, and the natural vegetation of bushy transitory forest has almost disappeared. Today, natural tree line habitats exist in very few places in the Andes such as in the almost inaccessible northern branch of the Cordillera Vilcabamba, 70-100 km northwest of Machu Picchu. It has been estimated that less than 1% of the potential cover of *Polylepis* forest remains on the humid eastern slopes of the Peruvian Andes. Concerted conservation efforts are urgently needed in this zone (Fjeldså & Kessler 1996).

THE IMPORTANCE OF MACHU PICCHU IN IHE CONSERVATION OF ANDEAN BIRDS

The conservation of key areas for biodiversity is often in conflict with poverty-driven pressures on nature and/or national development strategies. It is essential to establish a balance between conservation and land development wherever possible. Whether we choose to protect biodiversity in a network of reserves or by encouraging better land use, an informed approach is needed. The ecoregion approach has been used to identify large zones of general conservation concern in South America (Dinerstein *et al.* 1995). However, interesting local patterns, such as the aggregate of endemic species near Machu Picchu, are overlooked by this approach. An alternate approach ('the hard way') is through the compilation of thousands of distributional records that exist in museums, publications, reports and archives, such as the work done in recent years at the Zoological Museum in Copenhagen (Fjeldså & Rahbek 1997, 1999). Birds have been classified and charted more completely and precisely than any other group of organisms and are, thus, most suited for this approach.

Computer-based graphical tools exist today for fast, interactive handling of distributional data for large numbers of species. The digitized distribution data consists of range maps, where species are assumed to be continuously present between collecting points unless the absence of records from well-studied sites suggests otherwise. For species that seem to be genuinely rare or locally distributed, only confirmed records are used. Such databases facilitate an analysis of the whole of South America with a resolution of one geographical degree (1°) (Fjeldså & Rahbek 1997, 1999). For the Andes region, where the data is the most detailed, a resolution of 15 geographical minutes (15') is used (Fjeldså *et al.* 1999).

The eastern slopes of the tropical Andes, and the adjacent forelands, are the ornithologically richest areas in the world (at the spatial scale of 1° x 1°), with the highest number of species found at the equator, although an almost equal number of species are found in Cusco and in northern Bolivia.

Conservationists often rank areas by the number of recorded species. However, this is not necessarily the best approach to save species from extinction. If we decide to protect the 50 most species-rich 1° areas in South America, these areas would contain altogether 77% of the bird species found on the continent. However, most of these species are widespread and would be redundantly represented in several of the species-rich cells, and only 40% of those species actually threatened would be covered. If instead we use computer algorithms to compare the distribution of species, then analyze the pattern of geographical complementarity to identify a minimal number of areas which cover all species, then efficiency is greatly increased. Under such a system, the 50 top scoring areas in South America would cover 96% of all species and 85% of those which are currently threatened. It is also possible to prioritize existing reserve networks and take account of areas where conflicting interests abound.

In a conservation plan for the Andes region based on more focused data (15'x15' grid cells), a minimum of 28 areas would be needed to protect all the birds of Peru. The Machu Picchu area was the top scorer, with six species that are not covered in any other target area. Areas with four or five unique species in Peru are found (from the north) in the Huancabamba valley, the Utcubamba/Colán area, around Balsas in the upper Marañón valley, in the Carpish Mountains near Huánuco, in the lake Junín area, Bosque Zárate above Lima and the Ampay area near Abancay. Other areas are of less importance. Those who planned the existing reserves did not have access to this kind of data. Unfortunately, their focus on sparsely populated areas, where there are few conflicting interests, meant a bias towards areas with large numbers of widespread species. This has resulted in areas containing some of the rarest and most localized species not being protected.

Fortunately, the Machu Picchu area has been protected because of its archeological interest, and it has become an ornithological safe haven where many other elements of biodiversity have been conserved as well. As indicated above, the patterns of human settlement in Machu Picchu and the Vilcanota Valley may have been conditioned by the predictable, benign climate of the area, which accounts too for the presence of the many rare and endemic birds found in the area.

BIRDING MACHU PICCHU AND THE CUSCO REGION

by Huw Lloyd

Birdwatching in the Cordillera Vilcabamba, its near neighbor the Cordillera Vilcanota, and the surrounding Cusco landscapes will change you. Birdwatching here is something special, something unique, something that will call on you to return again and again. Few places in the world can generate such a range of emotions and experiences amongst birders in just a matter of days or weeks. There are the Vilcabamba humid cloud forests that carpet row-upon-row of knife-top ridges across the Machu Picchu landscape, forests that house some of the most spectacular and challenging mixed-species flocks. High above the Sacred Valley, tiny patches of threatened *Polylepis* woodlands are nestled beneath the snow line, providing critical resources for high-Andean people and homes for many of the regions' most unique birds. There are the *puna* grasslands, often overlooked en route to other destinations but perhaps one of the most important indicators of climate change. Remnant strips of semi-humid montane scrub survive on several slopes and serve as important habitat corridors for many woodland bird species. The thorny arid montane scrub habitats close to the outskirts of Cusco are home to range-restricted hummingbirds and *Furnariids*. Small bodies of wetlands adjacent to many towns and villages are important staging grounds for migrant waterbirds as well as hosting a wealth of resident species. The treeline elfin-forests of the Cordillera Vilcanota are the gateway for what is perhaps the most spellbinding and accessible gradient of Andean habitats for any birder along the entire eastern Andean slope.

Collectively this region constitutes an assemblage of one the most species rich bird communities in the world, and one of the largest irreplaceable areas for endemic birds in the Andes (Swenson *et al.* 2012). Why these areas are home to such biological riches has been the subject of much research in recent decades. What we do know is that the contemporary patterns of Andean bird species diversity we see today have been significantly shaped by the collective influence of many historical events.

ORIGINS AND EVOLUTION OF ANDEAN BIRD COMMUNITIES

Although the formation of the Andes began during the Triassic period 200 to 250 million years ago, the Andes only became truly isolated from the adjacent lowlands by the great Andean uplift, which began between 20 and 30 million years ago. This ancient uplift was not a constant progression, but different parts of the Andes were subject to different degrees of geological forces and uplift. During the Miocene epoch (6 to 10 million years ago), a pulse of rapid uplift caused parts of the central Andes to reach elevations between 1,500 m and 3,000 m, resulting in the complex mosaic of high mountains, deep inter-Andean valleys and layers of ridge-top foothills that birdwatchers traversing the Andes can see today.

This complex Andean topography became the most important continent-wide driver of the South American climate, influencing successive periods of climatic stability, and creating the humid conditions necessary to support such a wealth of bird diversity and endemism. Perhaps the topography's most profound impact was its role in regulating the severity of the south polar winds during the Pleistocene epoch some 11,000 to 2.5 million years ago, one of the most climatically turbulent periods in the Earth's history. These winds are a characteristic feature of the southern and central Andean climate, particularly in Peru and Bolivia, causing brief 10-day events of winter cold with occasional harsh winter freezes (Ronchail 1989). During the Pleistocene, the south polar winds were much more severe, resulting in drastic changes in montane vegetation. However, the Andes afforded topographical protection for many areas that may have acted as 'refugia' for many bird communities. In fact, recent DNA studies of Andean birds have revealed that most endemic and restricted-range species are relict populations of formerly widespread species that managed to survive in these humid refugia, where the Andean topography moderated the impacts of the Pleistocene climatic upheaval (Fjeldså *et al.* 1999).

The interaction of Andean topography and climate is just as apparent today. As humid air rises up the eastern slope of the Andes from the neighbouring Amazonian lowlands, it is cooled rapidly upon interacting with the Andean topography. Many Andean forested slopes are regularly enveloped in clouds, giving rise to the commonly used term 'cloud forests'. Concurrent with this is the remarkable 0.6°C variation in temperature with every 100 m change in elevation. This temperature range also influences the structure of Andean vegetation along the elevational gradient and the resultant patterns of forest-dependent bird communities (Still *et al.* 1999, Raxworthy *et al.* 2008). Today, the greatest numbers of endemic bird species are concentrated in these 'climatically stable, hyper-wet areas', where the cloud cover and mist formations are highly predictable (Long 1994, Fjeldså *et al.* 1999).

The origin and evolution of Andean wetland bird communities are also closely tied to Pleistocene climatic events (Fjeldså 1985). During this period in some regions of the high Andes, glaciers were present at 3,500 m elevation, even damning lakes around the Cusco region, (Fjeldså 1985). These repeating glacial and interglacial periods represented periods of extreme isolation and connectivity for wetland bird populations, with areas of tundra-like grasslands or glacial lakes such as Lago Junín in central Peru, remaining ice-free and forming 'refugia' for many wetland birds during the harshest glacial periods, thus giving rise to a number of unique species (Fjeldså 1985).

HUMID CLOUD FORESTS AND THEIR BIRD COMMUNITIES

Perhaps the most iconic image of Machu Picchu is that of the ruins surrounded by a blanket of humid cloud forest habitat covering the steep-sided slopes of the surrounding mountains. However, what seems to be a uniform forest habitat is actually a large mosaic of tightly interwoven humid cloud forests that vary in structure and species composition, depending on the precipitation gradient, cloud cover, soil composition and the steepness of the slopes. Landslides are frequent, forming distinctive scars on the slopes, which encourages a natural succession of shrubs and bamboo thickets.

Birdwatchers will find a variety of different terms used to classify humid cloud forests, based primarily on elevation and precipitation. At Machu Picchu the cloud forests between 2,500-3,400 m elevation are known as humid montane forests, whereas those <2,500 m are referred to as humid pre-montane forest habitats. Around the road and town of Aguas Calientes the 20-30 m tall pre-montane forest changes quite markedly in structure and species composition as you travel farther upwards in elevation toward the ruins of Machu Picchu at 2,600-2,700 m elevation. Here the most obvious features of the forest are the shorter, smooth, straight-trunked trees with compact and rounded crowns, with twisted branches that are covered in a profusion of mosses and epiphytic plants such as bromeliads, ferns, and orchids providing niches for numerous bird species (Stotz *et al.* 1996). Some slopes are dominated by fast-growing pioneer tree species such as *Cecropia* with their distinctive leaves and characteristic hanging clusters of inflorescences and seeds, which prove a magnet for mixed-flocks of birds (Stotz *et al.* 1996). The understory is dominated by various shrubs, melastomes, tree ferns, and *Chusquea* bamboo (Stotz *et al.* 1996) and is normally impenetrable unless punctured by roads or well-defined trails.

The bird communities of these humid cloud forest slopes are characterised by a remarkable vertical segregation of species that exhibit narrow, 'belt-like' elevational distribution, some of which may only be a few hundred meters wide (Stotz *et al.* 1996). From lower to higher elevations, many species similar in appearance and vocalisations, appear to 'replace each other' along the elevational gradient (Garcia-Moreno & Fjeldså 1999). Research from the Cordillera Vilcabamba suggested that this 'fine packing' of narrow elevational ranges was the result of interspecific competition –i.e. competition between individuals of different but closely related bird species (Terborgh 1971, 1977). This is somewhat understandable given the 'sharpness' of these vertical replacements and the closely related species involved. In actual fact, the patterns are not a consequence of competition or elevation *per se*, but are actually generated by multiple interacting abiotic and biotic factors operating at both local and landscape scales to influence the persistence of avian populations (Brown 2001, Holt & Keitt 2005). At the local scale, physical and ecological factors such as the local climate, ecotones, competition, habitat structure and complexity play a more prominent role (Terborgh 1985, Kessler *et al.* 2001). At the much wider landscape scale, biogeographic factors such as area of habitat, degree of isolation, and climatic gradients are some of the more important determinants of community composition.

ELFIN FOREST BIRD COMMUNITIES

The upper elevational forests that reach the Andean treeline ecotone above 3,200 m in Peru are known as elfin forests. These unique, stunted forests almost resemble something from a science fiction movie, being characterized by shorter, stunted trees that possess a remarkable flattened crown. Geographically, elfin forests form a near-continuous linear band throughout some areas of the high Andes, and are subject to highly precipitous and windy conditions on a daily basis. They play an important functional role in the high Andean ecosystem as a source of water for the whole montane basin during the dry season by using their distinctive foliage to condense moisture. Recent research from Peru has revealed that Andean treeline ecotones represent a distinctive vegetation community that can be easily differentiated from the adjacent cloud forest and *puna* grasslands based on changes in tree-size characteristics and vegetation cover (Lloyd *et al.* 2012), and that they may share a common spatial patterning and structural complexity throughout the Peruvian Andes (Young 1993).

Elfin forest bird species exhibit a particularly high level of disjunct ranges and strong differentiation between different populations (Graves 1985, 1988). This may be a consequence of the elfin forest's extremely narrow linear configuration, which may increase the likelihood of local extinctions and divergent evolution of localized bird populations (Graves 1985, 1988). Today, the composition of treeline bird communities is influenced by a pool of bird species from a much wider elevational gradient encompassing both cloud forest and *puna* grassland habitats (Lloyd *et al.* 2012).

PUNA GRASSLAND AND STEPPE BIRD COMMUNITIES

The often overlooked birds of the *puna* grasslands and neighboring steppe habitats are no less remarkable than their radically different humid cloud forest counterparts. Although there are far fewer habitat specialists (compared to other montane habitats) and far fewer species (Stotz *et al.* 1996), *puna* grassland bird communities show remarkable resilience to human disturbance, having been exposed to traditional agricultural practices over centuries (Fjeldså & Krabbe 1990). Today, many *puna* bird species have adapted and can be found near areas of human habitation. Typically, *puna* bird communities are dominated by common (abundant) species with wide geographic distributions (Fjeldså & Krabbe 1990). Next to nothing (empirically) is known regarding how *puna* bird communities are structured. *Puna* birds and those of the rocky, barren steppes that reach the snowline are less colorful, being typically a mixture of brown or grey colors. Species which occupy *Ichu* (*puna*) bunchgrass are secretive and hard to observe. In this zone, species tend not to move in mixed flocks, although several do wander in monotypic flocks. Many of these species are 'pioneers' and are adapted to breeding in this extreme environment, exposed to drought, high levels of UV, extremely variable weather conditions, and temperatures (Fjeldså & Krabbe 1990). How they separate ecologically from each other has yet to be resolved and is one of the most interesting issues in Andean avian ecology.

POLYLEPIS BIRD COMMUNITIES

Scattered throughout the Cordillera Vilcanota and Vilcabamba, at elevations above 3,400 m are small, highly fragmented areas of an ancient high-Andean woodland habitat with a bird community like no other. These relict *Polylepis* woodlands resemble something out of a science fiction movie, composed of reddish-brown trees which appear twisted and gnarled by the extremities of the high Andean climate. *Polylepis* bark is finely laminated, and the multiple layers are easily peeled off in the hand like sheets of paper. *Polylepis* trees can vary in height from 1.5 m upward of 20 m or more, depending on the species, elevation and level of human disturbance. They have characteristic small leaves, tightly arranged on the spiraling and twisting braches that are often coated with lichens. In areas where there is less livestock grazing, the sloping, uneven boulder-strewn ground beneath these woodlands is carpeted with mosses, a critical foraging resource for some *Polylepis* birds. Sadly woodlands of this quality are now only found in just a few areas in the Cusco region. Perhaps the best of these is Mantanay, situated at 3,800-4,600 m, high above the village of Yanahuara in the Cordillera Vilcanota, It is one of the few areas where *Polylepis* still forms a continuous elevational strip of woodland habitat with semi-humid montane scrub forest down toward the Sacred Valley.

Polylepis woodlands have provided essential natural resources and ecosystem services for rural Andean people, who represent some of the poorest rural communities in the Andean region, and who have utilized these woodlands for centuries. Historically, *Polylepis* woodlands were much more common and widespread throughout the high Andes (Fjeldså & Kessler 1996). Examination of fossil pollen records from Peru over the last ca. 370,000 years has revealed that the distribution of *Polylepis* woodlands has undergone dramatic fluctuations (Gosling *et al.* 2009). *Polylepis* did not always form a permanent continuous band of woodland habitat, even before the arrival of humans to the high Andes. Rather, *Polylepis* woodlands have shown to be extremely sensitive to past global climate change events, which have driven a series of rapid vegetation changes over thousands of years (Gosling *et al.* 2009). *Polylepis* woodlands were more extensive during warmer and wetter periods, but became much more restricted and fragmented during much drier periods (Gosling *et al.* 2009). Today however, there is little doubt that the present-day distribution of *Polylepis* woodlands in Peru and elsewhere within the Andes is a direct result of unsustainable human activity (Ellenberg 1958, Fjeldså & Kessler 1996).

Around half of all bird species that occupy *Polylepis* woodlands in the Cusco region are *Polylepis*-dependant and intolerant of the surrounding *puna* or steppe habitat (Lloyd & Marsden 2008). Most of these species occur at low densities in the tiny woodland patches that remain (Lloyd 2008a) and their population persistence is linked with both the quantity and quality of remaining woodland habitat (Lloyd 2008a, 2008b). The bird community is dominated by unique insectivorous species, many of which have very narrow niches and exhibit highly specialized foraging behaviors which enables them to coexist as they move through the woodlands in mixed flocks or smaller monotypic flocks (Lloyd 2008c). However some endemic species are able to shift their foraging behaviors to feed on reduced and more generalized prey resources in smaller woodland fragments (Lloyd 2008c).

ANDEAN WETLAND BIRD COMMUNITIES

The composition and structure of Andean wetland bird communities primarily reflects the vegetation and trophic conditions of the wetland ecosystem, rather than the quality of water (Fjeldså 1985, Stotz *et al.* 1996). Some wetland birds are able to feed directly on aquatic plants and, consequently, many herbivorous species are found here (Fjeldså 1985). However, wetland bird communities exhibit very limited endemism (Stotz *et al.* 1996), particularly the wetlands of the *puna* zone, which include lakes, marshes, highland bogs and wetter meadow areas with associated streams (Fjeldså 1985). Nevertheless, many species exhibit a number of physiological and behavioral adaptations to cope with the thin air, strong solar radiation, and exposure to temperature extremes on a daily basis (Fjeldså 1983). Most waterbird communities of the Cusco region are composed of widespread abundant species some of which are found in other Andean and lowland coastal wetlands, although there may be some differentiation between these populations (Fjeldså 1985).

Some of the most obvious vegetation characteristics of the larger high altitude Andean wetlands include tule and rush marshes (Fjeldså 1985). These areas have, like other high altitude Andean habitats of the Cusco region, been exploited for their vegetation and water for many centuries. Areas of marsh vegetation continue to be utilized and altered by rural communities today, creating more open water areas, helping to increase the number of waterbirds that are able to feed and breed in these wetlands (Fjeldså 1985).

BIRD CONSERVATION IN THE MACHU PICCHU/CUSCO REGION

Throughout the Peruvian Andes, anthropogenic impacts have been particularly severe on humid cloud forests, whilst the loss and degradation of *Polylepis* woodland is of particular concern because of its limited extent (only 2-3% of the former woodland cover remains in Peru), highly fragmented distribution, and the inadequacy of its protection within the national protected area network (Fjeldså 1993, 2002). The landscapes around Cusco are severely degraded due to frequent burning to create pasture for predominantly non-native livestock such as sheep and cattle. Natural, largely undisturbed treeline ecotones can only be found in some the most inaccessible regions of the Cordillera Vilcabamba (Walker & Fjeldså 2005). Despite the biological richness of the region, surprisingly little is known about the most basic ecological aspects of the bird communities of the Machu Picchu and Cusco region. With such a lack of empirical data, it is becoming increasingly difficult for conservation scientists to predict the consequences for these birds of continued habitat change or to address any potential problems associated with global climate change (Lloyd & Marsden 2008).

The establishment of national protected areas such as 326 km^2 Machu Picchu Historical Sanctuary is of critical importance for avian conservation. These areas may provide places to which bird species may retreat under adverse climatic conditions. However, efforts to conserve Andean birds must also encompass rural poverty-alleviation initiatives, focusing efforts at the interface between habitat conservation and natural resource use. Creating communal reserves such as those undertaken by the Peruvian non-governmental organization Asociación Ecosistemas Andinos (ECOAN) for rural Andean communities, may go some way to resolve land tenure conflicts whilst bringing knowledge of traditional land use practices into the heart of conservation efforts. These efforts must be supplemented by habitat restoration schemes that support other forms of sustainable land use as realistic alternatives to extensive pastoralism with non-native livestock (Fjeldså & Kessler 1996). Whilst some doubt that extensive forest ecosystems can be restored in such a manner across high Andean landscapes, evidence suggests otherwise. Research has shown that the Incas were able to implement landscape-wide agro-forestry schemes to restore ecosystem functions to the Cusco region which had become deforested and degraded some 1,000-4,000 years ago (Chepstow-Lusty *et al.* 1998). These systems were maintained for many years and only collapsed with the Spanish conquest (Chepstow-Lusty *et al.* 1998).

THE IMPACTS OF CLIMATE CHANGE ON ANDEAN BIRDS

Andean birds may be especially sensitive to the impacts of climate change (Feely & Silman 2010). Across the globe, montane forest habitats and species have migrated on average 6.1 m higher up mountain slopes per decade in response to increased temperatures (Parmesan & Yohe 2003). With global temperatures predicted to increase by 1.8 - 4°C by the year 2100 (IPCC 2007), humid montane forest bird species are predicted to shift their elevational range around 500 m upslope (Gasner *et al.* 2010). These migrations involve significant changes - an 'imbalance' or asymmetry - in the ratio of net colonization and extinction of populations at both the lower and upper limits of their elevational ranges, whilst the core of its elevational range becomes more fragmented. It is important to note that bird species are not directly dependent on temperature, but may be responding to gradual changes to the habitat, food availability, breeding resources and the presence of competitors (Forero-Medina *et al.* 2011). Not all birds are predicted to respond in a similar manner. The ability of a bird species to track climatic changes and reach new suitable habitat may be constrained by numerous factors, particularly if the species is restricted to a particular habitat or has a very narrow elevational range already impacted by habitat loss and degradation (Laurance *et al.* 2011).

The rapidly changing Andean climate will alter the strength and character of interactions between bird species and their environments. Rising temperatures in coming decades may signal changes to the Andean cloud base, which could shift hundreds of metres upward during the dry season, having dramatic effects on forest desiccation and precipitation (e.g. Still *et al.* 1999). Interspecific competition and differences in dispersal will dramatically affect bird community responses to climate change by decreasing abundances of species, increasing the risk of extinction and slowing their ability to track further climate change, and also preventing other species from colonizing newly available habitats (Urban *et al.* 2012). Extinctions caused by competition may be more likely for species inhabiting the highest-elevation habitats on mountaintops, such as the birds of the *puna* grasslands and *Polylepis* woodlands, which have much less 'ecological room for maneuver' in the face of any upward migration of species from lower elevations. Andean birds may become exposed to an array of new diseases, pathogens, and predators (Laurance *et al.* 2011). The net result is that climate-induced elevational shifts and biological interactions may bring together allopatric species, separate sympatric species, alter predator-prey relationships and in the process create a novel assemblage of species that currently does not exist (Williams *et al.* 2007, Urban *et al.* 2012).

Recent evidence suggests that along some Andean slopes, elevational range shifts are already occurring. Research from the eastern Andean slope near the Manu Biosphere Reserve in southeastern Peru show that Andean tree species are responding to climate change through direct upward elevational migration over 10 year periods (Feely *et al.* 2010). These responses are similar to the historical elevational migrations that occurred over much broader geological time-scales (Bush *et al.* 2004). A recent study from the Cerros del Sira in central Peru found an average upward elevational shift of almost 50 m for the majority of Andean bird species over a 41 year interval (Forero-Medina *et al.* 2011). This upward shift is actually much lower than the 152 m shift in average elevational range the scientists predicted using various climate change scenarios, meaning that there may be a time lag between climate change and the ability of Andean bird species to adequately respond. Conservation scientists, social scientists, ornithologists, policy makers, non-governmental organizations, and both regional and national governments must radically improve the way evidence-based conservation strategies are designed, funded and implemented, if they are to maintain biological interactions, create new or bolster existing protected areas, plant upslope habitat buffers, or succeed in alleviating Andean rural poverty.

BIBLIOGRAPHY

Bush, M. B., Silman, M. R. & Urrego, D. H. (2004). 48,000 years of climate and forest change in a biodiversity hot spot. *Science* 303(5659): 827-829.

Chepstow-Lusty, A. J., Bennett, K. D., Fjeldså, J., Kendall, A., Galiano, W. & Tupayachi Herrera, A. (1998). Tracing 4,000 years of environmental history in the Cuzco area, Peru, from the pollen record. *Mountain Research and Development* 18(2): 159-172.

Cracraft, J. (1985). Historical biogeography and patterns of differentiation within the South American avifauna: areas of endemism. - In: Buckley, P. A. *et al.* (eds). Neotropical ornithology. American Ornithologists Union, Lawrence, Kansas. pp. 49-84.

Ellenberg, H. (1958). Wald oder Steppe? Die naturliche Pflanzendecke der Anden Perus. *Umschau in Wissenschaft und Technik* 21: 645-681.

Feeley, K. J. & Silman, M. R. (2010). Land-use and climate change effects on population size and extinction risk of Andean plants. *Global Change Biology* 16: 3215-3222.

Fjeldså, J. (1985). Origin, evolution, and status of the avifauna of Andean wetlands. *Ornithological Monographs* 36: 85-112.

Fjeldså, J. (1992). Biogeographic patterns and evolution of the avifauna of relict high-altitude woodlands of the Andes. *Steenstrupia* 18: 9-62.

Fjeldså, J. (1993). The avifauna of the *Polylepis* woodlands of the Andean highlands: the efficiency of basing conservation priorities on patterns of endemism. *Bird Conservation International* **3**: 37-55.

Fjeldså, J. (1995). Geographical patterns of neoendemic and relict species of Andean forest birds: the significance of ecological stability areas. - In: Churchill, S. P. *et al.* (eds). Biodiversity and conservation of neotropical montane forests. New York Bot, Gard., New York, pp. 79-87.

Fjeldså, J. (2002a). Polylepis forests – vestiges of a vanishing ecosystem in the Andes. *Ecotropica* 8: 111-123.

Fjeldså, J. (2002b). Key areas for conserving the biodiversity of *Polylepis* forests. *Ecotropica* 8: 125-131.

Fjeldså, J. & Kessler, M. (1996). Conserving the biological diversity of *Polylepis* woodlands of the highland of Peru and Bolivia. A contribution to Sustainable Natural Resource Management in the Andes. Centre for Tropical Biodiversity and NORDECO, Copenhagen.

Fjeldså, J. & Krabbe, N. (1990). Birds of the High Andes. Zoological Museum, University of Copenhagen, Copenhagen, and Apollo Books, Svendborg.

Fjeldså, J., Lambin, E. & Mertens, B. (1999). Correlation between endemism and local ecoclimatic stability documented by comparing Andean bird distributions and remotely sensed land surface data. *Ecography* 22: 63-78.

Forero-Medina, G., Joppa, L., Pimm, S. L. (2010). Constraints to species' elevational range shifts as climate changes. *Conservation Biology* 25(1): 163-171.

Forero-Medina, G., Terborgh, J., Socolar, S. J. & Pimm, S. L. (2011). Elevational ranges of birds on a tropical montane gradient lag behind warming temperatures. *PLoS ONE* 6(12): e28535. doi:10.1371/journal.pone.0028535.

Garcia-Moreno, J. & Fjeldså, J. (1999) Re-evaluation of species limits in the genus Atlapetes based on mtDNA sequence data. *Ibis* 141: 199-207.

Gasner, M. R., Jankowski, J. E., Ciecka, A. L., Kyle, K. O. & Rabenold, K. N. (2010). Projecting the local impacts of climate change on a Central American montane avian community. *Biological Conservation* 143: 1250-1258.

Gosling, W. D., Hanselman, J. A., Knox, C., Valencia, B. G. & Bush, M. B. (2009), Long-term drivers of change in *Polylepis* woodland distribution in the central Andes. *Journal of Vegetation Science* 20: 1041-1052.

Graves, G. R. (1985) Elevational correlates of speciation and intraspecific geographical variation in plumage in Andean forest birds. *Auk* 102: 556-579.

Graves, G. R. (1988) Linearity of geographic range and its possible effect on the population structure of Andean birds. *Auk* 105: 47-52.

Holt, R. D. & Keitt, T. H. (2005). Species borders: a unifying theme in ecology. *Oikos* 108: 3-6.

IPCC. (2001). Intergovernmental Panel on Climate Change third assessment report – climate change 2001. IPCC, Geneva, Switzerland.

Kessler, M., Herzog, S. K., Fjeldså, J. & Bach, K. (2001). Species richness and endemism of plant and bird communities along two gradients of elevation, humidity and land use in the Bolivian Andes. *Diversity and Distributions* 7(1-2): 61-77.

Laurance, W. F., Camargo, J. L.C., Luizão, R. C. C., Laurance, S. G., Pimm, S. L., Bruna, E. M., Stouffer, P. C., Williamson, G. B., Benítez-Malvido, J., Vasconcelos, H. L., Van Houtan, K. S., Zartman, C. E., Boyle, S. A., Didham, R. K., Andrade, A. & Lovejoy, T. E. (2011). Global warming, elevational ranges and the vulnerability of tropical biota. *Biological Conservation* 144(1): 56-67.

Lloyd, H. (2008). Abundance and patterns of rarity of *Polylepis* birds in the Cordillera Vilcanota, southern Perú: implications for habitat management strategies. *Bird Conservation International* 18: 164-180.

Lloyd, H. (2008). Foraging ecology of high-Andean insectivorous birds in remnant *Polylepis* forest patches. *Wilson Journal of Ornithology* 120(3): 531-544.

Lloyd, H. (2008). Influence of within-patch habitat quality on Polylepis bird abundance. *Ibis* 150: 735-745.

Lloyd, H. & Marsden, S. J. (2008). Bird community variation across *Polylepis* woodland fragments and matrix habitats: implications for conservation within a high Andean landscape. *Biodiversity and Conservation* 17: 2645-2660.

Lloyd, H. & Marsden, S. J. (2011). Between-patch bird movements within a high-Andean *Polylepis* woodland/matrix landscape: implications for habitat restoration. *Restoration Ecology* 19(1): 74-82.

Lloyd, H., Sevillano Ríos, S., Marsden, S. J. & Valdez-Velásquez, A. (2012). Bird community composition and abundance across an Andean tree-line ecotone. *Austral Ecology* 37: 470-478.

Parmesan, C. & Yohe, G. (2003). A globally coherent fingerprint of climate change impacts across natural systems. *Nature* 421: 37-42.

Parmesan, C., Ryrholm, N., Stefanescu, C., Hillk, J. K., Thomas, C. D., Descimon, H., Huntley, B., Kaila, L., Kullberg, J., Tammaru, T., Tennent, W. J., Thomas, J. A. & Warren, M. (1999). Poleward shifts in geographical ranges of butterfly species associated with regional warming. *Nature* 399: 579-583.

Pounds, J. A., Fogden, M. P. L. & Campbell, J. H. (1999). Biological response to climate change on a tropical mountain. *Nature* 398: 611.

Raxworthy, C. J., Pearson, R. G., Rabibiso, N., Rakotondrazafy, A. M., Ramanamanjato, J., Raselimanana, A. P., Wu, S., Nussbaum, R. A. & Stone, D. A. (2008), Extinction vulnerability of tropical montane endemism from warming and upslope displacement: a preliminary appraisal for the highest massif in Madagascar. *Global Change Biology* 14: 1703-1720.

Ronchail, J. (1989). Advections polaires en Bolivie: mise en évidence et caractérisation des effets climatiques. *Hydrologie continentale* 4: 49-56.

Still, C. J., Foster, P. N. & Schneider, S. H. (1999). Simulating the effects of climate change on tropical montane cloud forests. *Nature* 398: 608-610.

Stotz, D. F., Fitzpatrick, J. W., Parker, T. A. & Moskovits, D. K. (1996). Neotropical birds: ecology and conservation. The University of Chicago Press, Chicago.

Swenson, J. J., Young, B. E., Beck, S., Comer, P., Córdova, J. H., Dyson, J., Embert, D., Encarnación, F., Ferreira, W., Franke, I., Grossman, D., Hernandez, P., Herzog, S. K., Josse, C., Navarro, G., Pacheco, V., Stein, B. A., Timaná, M., Tovar, A., Tovar, C., Vargas, J. & Zambrana-Torrelio, C. M. (2012). Plant and animal endemism in the eastern Andean slope: challenges to conservation. *BMC Ecology* 12(1): 1.

Terborgh, J. W. (1971). Distribution on environmental gradients: theory and a preliminary interpretation of distributional patterns in the avifauna of the Cordillera Vilcabamba, Peru. *Ecology* 52(1): 23-40.

Terborgh, J. W. (1977). Bird species diversity on an Andean elevational gradient. *Ecology* 58: 1007-1019.

Terborgh, J. W. (1985) The role of ecotones in the distribution of Andean birds. *Ecology* 66: 1237-1246.

Terborgh, J. W. & Weske, J. S. (1975). The role of competition in the distribution of Andean birds. *Ecology* 56: 562-576.

Urban, M. C., Tewksbury, J. J. & Sheldon, K. S. (2012). On a collision course: competition and dispersal differences create no-analogue communities and cause extinctions during climate change *Proceedings of the Royal Society B* 279(1735): 2072-2080.

Walther, G-R., Post, E,. Convey, P., Menzel, A., Parmesan, C., Beebee, T. J. C., Fromentin, J-M., Hoegh-Guldberg, O. & Bairlein, F. (2002). Ecological responses to recent climate change. *Nature* 416: 389-395.

Williams, W., Jackson, S. & Kutzbach, J. (2007). Projected distributions of novel and disappearing climates by 2100 AD. *Proceedings of the National Academy of Sciences United States of America* 104: 5738-5742.

Young, K. R. (1993a). Tropical timberlines: changes in forest structure and regeneration between two Peruvian timberline margins. *Arctic and Alpine Research* 25: 167-174.

HABITATS AND MICRO-HABITATS OF THE MACHU PICCHU AND CUSCO AREA

Rivers
Over 5 m wide and characterized by thickets and small trees, stands of willows (*Salix*) etc. Many species once restricted to these habitats now thrive in agricultural areas and man-made secondary growth.

Forest Edge and Secondary Growth
This is primarily a man-made habitat but also naturally caused by landslides. It is characterized by early secondary growth from grass-shrub associations to fast growing secondary forest, such as trees of the *Cecropia* genus.

Humid Montane Forest
A habitat characterized by a persistent, frequent or seasonal low-level cloud cover, usually at the canopy level. It is mature forest on Amazon facing mountain slopes. Trees are typically laden with arboreal epiphytes such as orchids, bromeliads, mosses and ferns. Many tree species are present and include *Clusia, Ocotea, Weinmannia* and Melastomes. Montane forest goes down to around an elevation of 2500 m.

Humid Pre-Montane Forest
Similar in composition and character to humid montane forest, humid pre-montane forest contains elements of flora, and is a transition between, lowland Amazonian terra firme forest and true humid montane forest and is generally below 2500 m elevation.

Elfin Forest
As the name suggests this is a low, dense, stunted forest that occurs at tree line or on the crests of ridges. Arboreal epiphytes are abundant.

Polylepis Woodland
Polylepis (*Rosaceae*), locally known Queñua, is a genus of low tree with flaking red bark that grows in fairly open groves at high altitude above normal tree line on steep, rocky slopes, and is usually surrounded by puna grassland. Other shrubs such as *Gnoxys* and *Brachyotum* are often associated with *Polylepis* as well as mistletoes and lichens. A small number of very rare bird species are restricted to this habitat.

Puna
This is seasonally dry cold grassland at the highest elevations in the Andes, known as puna. Cushion plants and tussocky bunch grass (*Stipaichu*) are the dominant plants. Many areas are overgrazed by llamas and alpacas. Rocky slopes have some heather-like composite brush such as *Braccharis*.

Andean Bogs
Poorly drained permanently damp locations found in the high Andes above treeline mostly in valley bottoms.

Freshwater Lakes and Ponds
Found throughout the Cusco highlands. Some glacial lakes have no vegetation and some have a fringe of marsh growth –e.g. *Typha* and *Shirpus*. There are no saline lakes in our area.

Semi-humid and Humid Montane Scrub
This is a broad category that includes a variety of plant associations. Present in humid ravines along streams with trees such as Alders (*Alnus*), *Jacaranda*, *Escalonia*, *Mysine*, *Bacharis*, *Schinus molle* (Pepper) and *Caeasalpina*. Not continuous humid forest.

Arid Montane Scrub
Similar to the above with a variety of plant associations but more xerophytic and dry and on open, arid mountain slopes. Vegetation includes *Cantua, Berbis, Tecoma, Barnadesia, Agave, Spartium* and cactus including prickly pear cactus.

Agricultural Areas
Fields planted with grains such as corn (maize) barley, wheat, quinoa and amaranth, beans and potatoes, sometimes with hedgerows of natural vegetation or planted exotic trees and bushes.

Freshwater Marshes
Marshes are areas with standing or very slow moving water filled with aquatic vegetation such as grasses, sedges and cattails.

MICRO HABITATS

Bamboo
Several species of birds are partial to or found almost exclusively in bamboo which is found patchily in the understory of montane forest (genus *Chusquea*), mainly in disturbed areas. Bamboo seeds at long intervals from several to 20 years and some bird species are nomadic foraging and breeding in seeding bamboo patches.

Tree fall Gaps
Tree falls in continuous forest create light gaps in the canopy which allows the growth of dense low vegetation often with many vines

Viny Tangles
Thick tangled vine growth found in tree fall gaps, or growing in columns up large tree trunks in montane forest. Several bird species specialize in this micro-habitat.

Streamsides
Borders of fast running montane streams with very wet vegetation, mosses and lichens. Rapids and waterfalls are common.

Talus Slopes
Scree is a collection of broken rock fragments at the base of crags, mountain cliffs, volcanoes or valley shoulders that has accumulated through periodic rock fall from adjacent cliff faces. Landforms associated with these materials are called talus slopes.

BIRD SPECIES OF SPECIAL INTEREST AND CONCERN IN THE AREA

Source: Birdlife International 2015

THREATENED SPECIES

Hooded Tinamou *Nothocercus nigrocapillus* **VULNERABLE**

Based on a model of future deforestation, this species has been listed as Vulnerable *Nothocercus nigrocapillus* is known from the eastern slopes of the Andes in north-central South America. The range of the nominate subspecies extends from central Peru to Bolivia. This species is expected to lose nearly one quarter of suitable habitat within its range over three generations (20 years) based on a model of Amazonian deforestation. Given the susceptibility of the species to hunting and/or trapping, it is therefore suspected to decline by ≥30% over three generations.

Taczanowski's Tinamou *Nothoprocta taczanowskii* **VULNERABLE**

This species is Vulnerable because it is known from few locations within a small range, where its apparently required habitat is subject to continuing degradation. Presumably, this is causing some population declines. *Nothoprocta taczanowskii* is uncommon and probably local on the eastern massifs of the Andes (in the upper parts of deep valleys intersecting the Cordillera Oriental and in inter-montane basins in the Cordillera Central) in Peru, and the adjacent La Paz department of Bolivia. The species faces intense hunting pressure in many parts of its range, especially where it occurs in proximity to human habitation. It is also negatively affected by the burning of pampas grassland. It inhabits mosaics of cloud forest (*Podocarpus*, *Eugenia*, *Escallonia*, *Polylepis*), scrub, pastures, fields, open rocky or grassy areas, mainly in humid or semi-humid montane areas such as at Peñas below Abra Malaga.

Black-and-chestnut Eagle *Spizaetus isidori* **ENDANGERED**

This species is considered Endangered as it has a small population, with all sub-populations believed to number fewer than 1,000 mature individuals. Its numbers are continuing to decline as a result of the destruction of its montane forest habitat as well as direct human persecution. Although it may persist in mosaics of primary and secondary forest with open areas, given habitat loss (Thiollay 1994) and persecution by humans (H. Vargas *in litt.* 2012) throughout its range, the population is considered to be declining. It is found on heavily forested mountain slopes, probably occurring mostly in large valleys, such as at Machu Picchu

Golden-plumed Parakeet *Leptosittaca branickii* **VULNERABLE**

Very high levels of forest clearance, fragmentation and degradation have presumably resulted in this species undergoing rapid population declines, qualifying it as Vulnerable. Total numbers are difficult to assess, but the population may be small. *Leptosittaca branickii* is widely but locally distributed and may now be declining in Peru (where its population had generally been considered to be stable) due to increasing habitat destruction (H. Lloyd *in litt.* 2007). A rapid and ongoing population decline is suspected on the basis of large-scale habitat destruction, degradation and fragmentation.

Royal Cinclodes *Cinclodes aricomae* **CRITICALLY ENDANGERED**

This species qualifies as Critically Endangered because its extremely small population is restricted to a severely fragmented and rapidly declining habitat (*Polylepis* forest), from which equivalent declines in population size are likely. Furthermore, all subpopulations are thought to be tiny. *Cinclodes aricomae* occurs in the Andes of southeastern Peru (Cusco, Apurímac, Puno, Ayacucho and Junín) and adjacent La Paz, Bolivia (C. Aucca Chutas *in litt.* 2012). A range-wide conservation plan for the species that was being drafted in 2010 estimated the total population at 231-281 individuals (D. Lebbin *in litt.* 2010), thus there are likely to be fewer than 250 mature individuals.

White-browed Tit-Spinetai *Leptasthenura xenothorax* **ENDANGERED**

This species has a very small and severely fragmented range and population in fragmented *Polylepis* forest, which continues to decline with habitat loss and a lack of habitat regeneration (Collar *et al.* 1992). Therefore, it is listed as Endangered. *Leptasthenura xenothorax* has a very restricted and severely fragmented range in the Runtacocha highland (Apurímac), the Nevado Sacsarayoc massif and the Cordillera Vilcanota (Cusco), south-central Peru. This species is suspected to lose more than half of suitable habitat within its range over three generations (11 years) based on a model of Amazonian deforestation (Soares-Filho *et al.* 2006, Bird *et al.* 2011). Given the susceptibility of the species to fragmentation and/or edge effects, it is therefore suspected to decline by ≥50% over three generations.

Puna Thistletail   *Asthenes helleri*   VULNERABLE
Based on a model of future deforestation, it is suspected that the population of this species will decline rapidly over the next three generations, and it is therefore listed as Vulnerable. *Asthenes helleri* has a restricted range in the Andes of western South America, described as uncommon to fairly common throughout. In Peru it is limited to the areas of Cusco and Puno, and is present in the Machu Picchu Historical Sanctuary. Its range also touches into extreme north La Paz, Bolivia. This species is suspected to lose about 30% of suitable habitat within its range over three generations (11 years) based on a model of Amazonian deforestation.

Marcapata Spinetail   *Cranioleuca marcapatae*   VULNERABLE
Based on a model of future deforestation in the Amazon basin, and its dependence on primary forest, it is suspected that the population of this species will decline rapidly over the next three generations, and it has therefore been up listed to Vulnerable. *Cranioleuca marcapatae* is present in the Machu Picchu Historical Sanctuary. It is generally uncommon. This species is suspected to lose over 20% of suitable habitat within its range over three generations (11 years) based on a model of Amazonian deforestation (Soares-Filho *et al.* 2006, Bird *et al.* 2011). Given the susceptibility of the species to fragmentation and/or edge effects, it is therefore suspected to decline by ≥30% over three generations.

Ash-breasted Tit-Tyrant   *Anairetes alpines*   ENDANGERED
This species has a very small population and is confined to a habitat (*Polylepis* woodland) which is severely fragmented and undergoing a continuing decline in extent, area, and quality. It is consequently listed as Endangered. *Anairetes alpinus* occurs locally in the high Andes of Peru and Bolivia. It is relatively common in the Runtacocha highland, Apurímac, and the Cordillera Vilcabamba, Cusco (Fjeldså and Kessler 1996), with the population at Abra Malaga estimated at c.20-30 birds (Engblom *et al.* 2002). In Bolivia it is locally common at the north end of the Cordillera Real in the Cordillera Apolobamba, and the total Bolivian population was estimated at 150-300 birds in 2007 (I. Gomez *in litt.* 2003, 2007). The total population is perhaps in the mid or upper hundreds. This species' population is suspected to be experiencing a moderate and ongoing decline, in line with habitat loss and degradation within its range.

White-tailed Shrike-Tyrant   *Agriornis albicauda*   VULNERABLE
This species is poorly known. It appears to be very rare to rare and very local throughout its range. Collar *et al.* (1992) described it as exceedingly rare. Given this, the total population is estimated to fall below 10,000 individuals, despite its large range. Trends have not been well documented, but the species appears to be declining for poorly understood reasons (B. Knapton *in litt.* 2003), being scarce even in areas where it was formerly described as relatively numerous (e.g. Ridgely and Tudor 1994). In Ecuador there has been 'an apparently precipitous drop in numbers' (Ridgely and Greenfield 2001).

Lemon-browed Flycatcher   *Conopias cinchoneti*   VULNERABLE
Based on a model of future deforestation in the Amazon basin, it is suspected that the population of this species will decline rapidly over the next three generations, and is therefore considered Vulnerable. *Conopias cinchoneti* has a disjunct range in the Andes of northwest South America (del Hoyo *et al.* 2004). The global population has not been quantified, but this species is described as 'fairly common but patchily distributed' (Stotz *et al.* 1996). This species is suspected to lose around 30% of suitable habitat within its range over three generations (11 years).

Cerulean Warbler   *Setophaga ceruleaa*   VULNERABLE
Setophaga cerulea breeds from Quebec and Ontario Canada, east to Nebraska and south to northern Texas, Louisiana, Mississippi, Alabama and Georgia USA (A.O.U. 1983) and winters from Colombia and Venezuela south, mainly east of the Andes, to eastern Ecuador, southeastern Peru and perhaps occasionally to northern Bolivia. This species has undergone a large and statistically significant decrease

over the last 40 years in North America (over 80% decline over 40 years, equating to over 35% decline per decade).

SPECIES OF LESS CONCERN BUT OF INTEREST DUE TO THEIR SMALL AND LOCAL POPULATIONS

Andean Condor   *Vultur gryphus*
Widespread in the Andes, increasingly rare in the north, common in the south especially Patagonia, increasingly declining at an unknown rate in Peru and more and more restricted to isolated areas. Still present in the Machu Picchu area and at Abra Malaga.

Semicollared Hawk   *Accipiter collaris*
Rare hawk of humid montane forest in southern Peru but seen often at Machu Picchu.

Imperial Snipe   *Gallinago imperialis*
Rare and local at tree line in elfin forest and *sphagnum* bogs. Can be seen or heard near Canchaillo at Abra Malaga.

Bearded Mountaineer   *Oreonympha nobilis*
Endemic to the central Peruvian Andes on scrubby slopes and in canyons. Huacarpay Lakes is one of the best places to see it. Look for its food plant *Nicotiana* (Tree Tobacco).

White-tufted Sunbeam   *Aglaeactis castelnaudii*
Endemic to the central Peruvian Andes, in semi-humid montane forest. Easily seen at Abra Malaga and on the first days of the Inca Trail hike.

Green-and-white Hummingbird   *Amazilia viridicauda*
An endemic to central-southern Peru in humid forest. Very common in the Urubamba Valley at Machu Picchu.

Orange-breasted Falcon   *Falco deiroleucus*
In southern Peru restricted to foothills at the base of the Andes. Rare – can be seen near the Machu Picchu Ruins.

Red-and-white Antpitta   *Grallaria erythroleuca*
An uncommon endemic in humid forest in south-central Peru. Can be seen near San Luis at Abra Malaga.

Rufous Antpitta   *Grallaria rufula occobambae*
A revision of this species will indicate that there are several species within the 'Rufous Antpitta' complex and the *occobambae* race will become a species restricted to southern Peru and north Bolivia. Has a two note call. Common near tree line at Abra Malaga and along the Inca Trail.

Vilcabamba Tapaculo   *Scytalopus urubambae*
A Peruvian endemic found only in the Cordillera Vilcabamba in Cusco. Common along the Inca Trail above 3500 m in shrubby forest and edge.

Diademed Tapaculo   *Scytalopus schulenbergi*
Found in stunted vegetation at tree line in southern Peru and northern Bolivia. Fairly easily seen at Abra Malaga.

Tawny Tit-Spinetail   *Leptasthenura yanacensis*
Found in Peru and Bolivia and restricted to groves of endangered *Polylepis* forest such as at Abra Malaga.

Line-fronted Canastero   *Asthenes urubambensis*
Local in Peru and Bolivia in humid tree line habitat and *Polylepis* groves.

Junín Canastero *Asthenes virgata*
Endemic to the central Peruvian Andes in bunch grass with scattered bushes. Can be found in the Cordillera Vilcanota and at Abra Malaga.

Scribble-tailed Canastero *Asthenes maculicauda*
Similar to the preceding species with the same habitat preferences and a similar song. Locally overlaps in our area in the Cordillera Vilcanota.

Rusty-fronted Canastero *Asthenes ottonis*
Endemic to the central Peruvian Andes – dry bushy slopes in intermontane valleys. Common around Huacarpay Lakes.

Creamy-crested Spinetail *Cranioleuca albicapilla*
Endemic to the central Peruvian Andes in semi-humid montane forest and shrubbery. Noisy. Can be seen along the first part of the Inca Trail and at Abra Malaga above Peñas.

Sclater's Tyrannulet *Phyllomyias sclateri*
Rare along east slope of the Andes in southern Peru and northern Bolivia. Quite common along he Urubamba River at Machu Picchu.

Inca Flycatcher *Leptopogon taczanowskii*
Endemic south of the Maranon River to Cusco and found in humid pre-montane forest.

Olive-sided Flycatcher *Contopus cooperi*
An uncommon boreal migrant. The species has undergone a moderately rapid decline and therefore qualifies as Near Threatened.

Bolivian Tyrannulet *Zimmerius bolivianus*
Found from Cusco to Bolivia in humid montane forest such as at Machu Picchu and Abra Malaga.

Masked Fruiteater *Pipreola pulchra*
Endemic to Peru, along the east slope of the Andes, and fairly common at Machu Picchu in the Urubamba Valley.

Inca Wren *Pheugopedius eisenmanni*
Endemic discovered at Machu Picchu and common in bamboo thickets around the ruins and at Abra Malaga.

Fulvous Wren *Cinnycerthia fulva*
Local from Cusco to Bolivia in humid montane shrubbery.

Slaty Tanager *Creurgops dentata*
Range restricted from Cusco to Bolivia in sub-montane forest like that at Machu Picchu.

White-browed Hemispingus *Hemispingus auricularis*
By some considered part of Black-capped Hemispingus but genetic differences indicate it is a separate species endemic to Peru. Found in humid montane forest at Abra Malaga.

Parodi's Hemispingus *Hemispingus parodii*
Found in Cusco Department in bamboo thickets in elfin forest. Fairly common at Abra Malaga.

Yellow-scarfed Tanager *Iridosornis reinhardti*
Distributed from northern Peru south to Cusco in montane and elfin forests. At the southern limit of its range in our area.

White-browed Conebill *Conirostrum ferrugineiventre*
Central Peru to Bolivia in elfin and *Polylepis* forest. Local.

Giant Conebill *Oreomanes fraseri*
Patchily distributed from southern Colombia to Bolivia and local in *Polylepis* woodland. Can be found at Abra Malaga.

Chestnut-breasted Mountain-Finch *Poospiza Caesar*
Peruvian endemic found in the Departments of Apurimac, Ayacucho and Cusco on bushy semihumid slopes.

Apurimac Brush Finch *Atlapetes forbesi*
Endemic to south-central Peru and rare in our area. Has been seen near Peñas and could be seen around the Salkantay massif.

Cuzco Brush Finch *Atlapetes canigenis*
Peruvian endemic restricted to Cusco in humid forest and shrubbery. Can be seen above Machu Picchu ruins near Intipunku and the lower areas of Abra Malaga above San Luis.

SPECIES ACCOUNTS

TINAMOUS Tinamidae

Tinamous are a strictly neotropical family. They are plump, slender-necked, small-headed birds with short wings and tails. They are terrestrial and furtive and hide by crouching and sitting still, only flushing in an explosive manner when almost stepped on. In the mountainous areas, such as the Andes, they prefer to fly downhill as they are weak fliers. Tinamous eat seeds, roots, insects and leaves. Females are larger and more aggressive. Most highland species are polygamous with two or more females laying eggs in the same scrape. Some species are polyandrous. Eggs are unicolored with a porcelain-like gloss.

Hooded Tinamou   *Nothocercus nigrocapillus*

33 cm. < 3000 m. Here the nominate *nigrocapillus*. Reddish brown, with dense fine black vermiculations producing a very dark effect. Inhabits humid montane and sub-montane forest under growth, bamboo stands (where it may concentrate when seeding) and dark places. Terrestrial, very timid and hard to see. Forages on the ground in search of seeds. Usually found alone or in pairs and often only reveals its presence by its far-carrying single-note calls *'brau brau'* repeated for long periods. Uncommon at Machu Picchu.

Brown Tinamou   *Crypturellus obsoletus*

26 cm. > 3000 m. Andean populations dark. Plumage deep rufous-chestnut with contrasting gray head and neck. See Hooded Tinamou. Inhabits rather open humid montane and sub-montane forest, forest edge, alder groves and mature secondary growth. Usually encountered walking quietly and feeding on the ground underneath tall forest, often along trails and the edge of narrow tracks. Feeds on seeds and fallen fruits. The song is a loud manic, rolling series of notes *'trehyrr-ree-ree-reee-ree,'* etc., accelerating consistently. The call is a loud *'tree-dreee,'* given every 12-15 seconds. Can be heard and seen on the slopes of Wayna Picchu near Machu Picchu ruins.

Taczanowski's Tinamou   *Nothoproctao taczanowskii*

36 cm. 2800-4000 m. Large and gray. Long curved bill and yellow legs. Looks very dark gray-brown with pale streaks and spots. Inhabits rocky and grassy slopes with some shrubbery, or scattered *Polylepis*. Also mosaics of partly-cleared or cultivated parts of montane shrub forest and the edges of fields, edges of potato crops. Terrestrial and runs rapidly along field edges and clearings. Difficult to flush. Uncommon, but frequently seen in and around Peñas on the west slope of Abra Malaga. Song a loud ringing whistle *'tu'eeeeeeer.'* This species is named for Władysław Taczanowski (1819-1890), Polish zoologist who wrote *Ornithologie de Pérou*, an early and comprehensive account of Peru's birds.

Ornate Tinamou   *Nothoprocta ornata*

35 cm. > 3000 m. A largish pale grayish-brown Tinamou of high, dry grasslands and bunch grass, sometimes on bushy, rocky slopes – Puna grassland. Note bushy crest and spotted head and neck and unspotted breast. Uncommon in our area and only likely to be seen at the higher elevations of the Sacred Valley of the Incas and around the Salcantay Massif. Flushes explosively and sits tight until the last minute. Song when flushed is a *'weechu-weechu-weechu-weechu'* until it plunges into cover. Also hen-like clucks on ground. Rare.

Andean Tinamou   *Nothoprocta pentlandii*

28 cm. 1800-3600 m. The *fulvescens* race is present in our area inhabiting intermontane valleys, a variety of drier habitats including thickets and ravines in semi-arid areas, bushy slopes with scattered trees and hillsides with scrub – *Lupinus* sp., montane scrub admixed with cactus and agricultural fields and edges of *polylepis* woodland. Head with dark spots on the crown, but note unspotted sides of face and neck and unspotted breast. Walks looking for food under the cover of dense bushes and crouches when danger approaches, only flushing at the last minute. Calls include a liquid *'yoo-tuu'* and when flushed a series of melodic notes *'pucyuu-pyucuu-pucc-pucc.'* Can be seen near the start of the Inca Trail at Llactapata and the Sacred Valley of the Incas including Huacarpay Lakes. Fairly common.

DUCKS AND GEESE   Anatidae

A worldwide family with little morphological variation between species. All have rather short tails and short legs with webbed feet for swimming and broad bills for sieving. Plumage is dense with a smooth surface. Most species show an iridescent patch or speculum on the secondaries. Ducks are awkward on the ground. They are flightless for about a month after breeding, when the wing and tail feathers are molted simultaneously. Ducks may be divided into two main groups – "diving ducks" and "dabbling ducks," which are surface feeders. Males are generally brightly-colored while females are drab. Some males attain a female-like plumage when molting (eclipse). Nests are simple scrapes lined with the female's own down. The female incubates alone.

Black-bellied Whistling Duck   *Dendrocygna autumnalis*
48-53 cm. 3000-3800 m. A distinctively shaped duck. Reddish bill and legs. Mostly rufous brown with a black belly. Found on lake margins and sometimes river banks. As its name suggests makes a reedy whistle. A rare visitor to lakes in our area with records from Lago Huaypo near Chinchero and, historically, Huacarpay lakes south of Cusco.

Andean Goose   *Oressochen melanopterus*
70-90 cm. > 3800 m (occ. 3200 m). Male largest. Both sexes mainly white with rosy-colored bill and feet. Found in open terrain with short grass, on bogs in wet valleys and around lakes and ponds. Usually encountered in loose flocks or pairs, large dense flocks when molting. Feeds by walking slowly with its head down, cropping succulent semi-aquatic plants. Has an elaborate strutting display in the breeding season. Flies well, frequently covering long distances between feeding areas, sometimes making a soft *'quip-quip'* call. Found in higher parts of the Sacred Valley and Abra Malaga, lakes Huaypo and Piuray (sometimes large numbers) and occasionally Huacarpay lakes in winter.

Torrent Duck   *Merganetta armata*
40 cm. > 1800 m. Red bill and long stiff tail. Male: black body, head and neck, white with black lines. Female: entire under-parts from bill to undertail chestnut. Inhabits clear boulder-strewn streams and rivers with rushing torrents interspersed with calm stretches, often in canyons and gorges. Usually found in pairs, sometimes in family groups. Pairs defend their stretches of river energetically. Often seen sitting upright on rocks and boulders among rushing rapids. Swims upstream in rapid white water, but mostly feeds in eddies, diving under water or dabbling at water's edge. Escapes danger by swimming downstream low in the water. Rarely flies. A characteristic bird of the Urubamba River, can be easily seen from the train to Machu Picchu where the railway line runs parallel to the river.

Crested Duck   *Lophonetta specularioides*
60 cm. > 3500 m. The *alticola* subspecies is present. A large long-bodied duck. Bill bluish. Mostly gray brown with a dark tail. Shows hanging crest. Primaries black. Inhabits high altitude lakes, lagoons, tarns and ponds but shows a preference for lakes with barren shores. Fairly common on high lakes in the Sacred Valley and occasionally at Huacarpay lakes near Cusco, also Piuray/Huaypo lakes near Chinchero.

Andean Duck   *Oxyura ferruginea*
45 cm. > 3000 m. Considered by some to be only a subspecies of Ruddy Duck *(Oxyura jamaicensis)*, as the subspecies *andina* of the Colombian Andes may be a population of hybrid origin. A stocky duck with a large shovel shaped bill. Male: bill cobalt blue, head and upper hind-neck black with (occasionally) white feathers on the cheeks, neck and body chestnut with blackish rump and tail. Female: gray bill, mostly dark brown with some mottling, lighter area below the eye and on the chin. Found on fairly deep clear lakes with water-weed. Usually in loose flocks away from other ducks, often sleeping in the center of lakes with the head buried in the wing. Unusually it holds tail at a cocked angle above the water. Dives for underwater vegetation for up to 30 seconds at a time. Mostly above 3000 m. Common on high lakes in the Sacred Valley of the Incas, lakes near Chinchero and Huacarpay lakes.

Yellow-billed Teal   *Anas flavirostris*
40 cm. 2500-4500 m. The *oxyptera* subspecies is found in our area. Jaramillo (2003) suggested that the subspecies *oxyptera* may deserve recognition as a separate species from *A. flavirostris* found in Patagonia. A compact short-necked duck. Sexes resemble each other. Bill yellow with a black ridge. Dark-hooded effect caused by dark, dense stippling to the head and upper neck. Inhabits all kinds of watery habitats from rivers and streams to ponds and boggy Andean valleys. In pairs or small groups often with Puna Teal or Yellow-billed Pintail. Often feeds out of the water on river and pond banks. Has a swift and erratic flight. Common at Piuray, Huaypo and Huacarpay lakes near Cusco.

Yellow-billed Pintail   *Anas georgica*
54-57 cm. > 3000 m. A slim duck with a slender bill, thin neck, and sharply pointed tail. Sexes resemble each other. Yellow bill with a black ridge. At a distance it looks uniformly buffy with a light-colored head. Inhabits wetlands, especially lakes with shallows with submerged or floating water-weeds. In pairs or flocks, often with other waterfowl. Walks well and often seen on lake shores feeding. Feeds in the water, usually by upending itself, but will make shallow dives. Mostly found above 3500 m. Fairly common at Piuray and Huaypo lakes near Chinchero. Uncommon, but seemingly resident at Huacarpay lakes south of Cusco.

White-cheeked Pintail   *Anas bahamensis*
44-47 cm. 3000 m. Shape similar to more common Yellow-billed Pintail. Bill blue-gray with black ridge and red spot at base. Brown capped with white face and throat. Body tawny with black spots. Feeds off floating vegetation, sometimes upending. Only known from Huacarpay lakes south of Cusco at 3000 m, where it is an uncommon (increasing?) visitor to lakes in the Cusco area.

Puna Teal   *Anas puna*
47 cm. > 3000 m. Longish straight blue bill. Black-capped to the level of the eyes with white cheeks and throat. Sides densely barred black and white (male) or more buffy (female). Rear of body pale gray and faintly mottled. Found on open lakes with submerged or floating vegetation and with islands or floating reed-beds which it uses for nesting. Usually encountered in dispersed flocks. Feeds off floating vegetation, sometimes upending. Common on high lakes in the Sacred Valley such as Piuray and Huaypo near Chinchero and at Huacarpay lakes south of Cusco.

Blue-winged Teal   *Anas discors*
37-41 cm. 3000 m. Uncommon boreal migrant. Male has head dull blue-gray with large white crescent in front of the eye. Female shows distinct whitish loral spot and dark eye line. Habitat similar to Cinnamon Teal. Recorded so far at Huacarpay, Huaypo and Piuray lakes near Cusco October-April, but undoubtedly occurs on other water bodies during the austral summer.

Cinnamon Teal   *Anas cyanoptera*
40-45 cm. > 3000 m. Long, somewhat spatulate gray bill. Male: chestnut, back feathers black with rusty edges. Uniform chestnut below, sometimes with some black spots. Female and eclipse male: buffy with a cinnamon tinge, obscurely mottled and spotted darker. Both sexes show a large pale blue forewing area and white and green speculum. Found in lakes, ponds and marshes with some reeds and floating aquatic vegetation. Social and gregarious but usually not found in large groups. Often with other ducks. Common on lakes in higher parts of the Sacred Valley and Huacarpay lakes south of Cusco.

Red Shoveler   *Anas platalea*
46-51 cm. 3000 m. Note large spatulate bill. Male with black bill, head pinkish buff stippled with black and cinnamon colored body. Female very similar to female Cinnamon Teal but recognized by bill shape. Distinctive habit of holding the head low running the bill along the surface of weedy and shallow muddy water. Found on lakes and marshes with an abundance of water weeds on the surface. In our area usually found alone or in pairs. Possibly only a rare austral migrant, but some may stay throughout the summer months. Recorded on several occasions from Huacarpay lakes south of Cusco.

Yellow-billed Teal

Yellow-billed Pintail

White-cheeked Pintail

Puna Teal

Blue-winged Teal

Cinnamon Teal

Red Shoveler

GUANS, CURRASOWS AND ALLIES Cracidae

Cracids are found exclusively in the New World. They have large, strong feet and legs and chicken-like bills. Wings are short and rounded. They live in pairs or small groups and feed on buds, fruits and flowers. They are prized for food and are extensively hunted. Guans give a characteristic wing – whirring display at dawn in the breeding season. The nests are simple twig and leaf platforms in a bush or tree and they lay three creamy eggs. The young are precocious and can climb along branches after 4-5 days.

Andean Guan *Penelope montagnii*
60 cm. 1800-3500 m. The *plumosa* subspecies is present. A large turkey-like bird, with an erectile crest. Bare area around the eye blue-gray. Small orange dewlap. Inhabits humid montane and pre-montane forest where there is a profusion of bromeliads. Small groups move silently through the canopy and sub-canopy of the forest in search of fruits etc. The song is a repeated *'chaah-choah-cha-cah-choam cha-cha-cha,'* etc., not often heard but given at dawn and dusk. In aerial display gives a whistle followed by muffled wing drumming. Can be seen in forested areas throughout the Machu Picchu Sanctuary. Named for Jean Pierre Francoise Camille Montagne, French surgeon and botanist (1784-1866). In the days before, during and after the Napoleonic Wars, many "Natural Philosophers" (early naturalists) including Charles Darwin joined naval expeditions during an era when new species were being discovered around the globe.

Sickle-winged Guan *Chamaepetes goudotii*
65 cm. 1800-2500 m. A more slender bird than the preceding species with longer legs. Does not show a crest or dewlap but the facial skin is light cobalt blue. Found in humid pre-montane forest where there are tall trees and precipitous slopes. Usually in pairs or small groups in the forest canopy, feeding in fruiting trees at dawn and dusk. Wary and elusive. Makes some clucking noises and a loud *'kee-uck'* when alarmed. The specialized, narrow, sickle shaped outer primaries make a rattle during the short display flight. Rare at Machu Picchu but has been seen along the upper part of the road just below the Machu Picchu ruins.

NEW WORLD QUAIL Phasianidae

In the Americas, the toothed quail sub-family is present. They are small compact birds with short very stout bills, rounded wings and short tails. They are terrestrial and live under cover of vegetation. Found in pairs or small family groups on the forest floor in dense undergrowth. When alarmed they run away quickly or freeze. Sometimes they cross trails and openings in single file. They are mostly detected by far-carrying whistled songs. The nest is a domed construction. The young are precocious and can flutter away within a week of hatching.

Rufous-breasted Wood-Quail *Odontophorus speciosus*
28 cm. 1800-2600 m. Stocky and thick-billed. Male: breast and under-parts mostly chestnut. Female: like male but underparts dark gray with a narrow chestnut band across the breast. Inhabits dense humid montane and pre-montane forest. The song is a complex series of liquid, far-carrying notes given by several birds in the group. Can be heard, rarely seen, at Machu Picchu ruins.

Striped-faced Wood-Quail *Odontophorus balliviani*
26 cm. 1800-3300 m. Stocky and thick-billed. Mostly chestnut brown with slight vermiculations. Underparts fulvous with diamond shaped white spots. Inhabits humid montane and pre-montane forest with epiphytes, bamboo and tree ferns. The song is a complex series of liquid notes often repeated – *'wheddi-dee wheedley-dee'* Uncommon, may be overlooked. Named for General José Ballivián y Segurola, President of Bolivia 1841-1847.

Andean Guan

Sickle-winged Guan

Rufous-breasted Wood-Quail

Striped-faced Wood-Quail

GREBES Podicipedidae

Grebes are foot propelled diving birds recognized by their pointed bills and downy, almost tail-less rear. The feet are set far back, not suitable for walking, but excellent for swimming. Each toe has a separate swimming lobe. Grebes are mainly aquatic, only leaving the water to fly to other bodies of water or to climb onto their nests, which are soggy floating structures among water plants. Andean grebes lay two eggs and the young can dive almost immediately upon hatching. The diet consists of small fish, crustaceans and insects which are gleaned from water weeds.

White-tufted Grebe   *Rollandia rolland*
24-30 cm. 3000-4500 m. Toward the end of the breeding season, the black areas become brown and underparts clearer. The throat and belly almost white and neck dull rufous in non-breeders. Inhabits marshes, ponds and shallow lakes and seems to prefer a mosaic of clear channels and pools admixed with aquatic vegetation. Rarely flies (when it does it shows large white patches in the wings). Shows high sterned profile with fluffy plumage. Territorial or loosely social. Often seen chasing one another, pattering across open water. Feeds on the surface but also dives for food. Prefers fish but will take aquatic insects. Fairly common on suitable bodies of water in the area. Named for Master Gunner Rolland. The species was first collected in the Falklands during the French Navy's circumnavigation of the globe 1817-1820.

Silvery Grebe   *Podiceps occipitalis*
27 cm. 3000-5000 m. Here the race *juninensis*. Fjeldså & Krabbe (1990) and Jaramillo (2003) suggested that the northern Andean subspecies, *juninensis*, might merit recognition as a separate species from *Podiceps occipitalis* farther south. Note the small bill. In flight shows gray inner primaries and secondaries. Inhabits open lakes with or without vegetation. Not skulking and usually highly visible in pairs, groups or dispersed flocks. Often sunbathes with fluffy plumage. Has an elaborate mating display which involves parallel races in upright attitudes with their bodies almost out of the water. Feeds mostly on insects by diving or picking them off the surface of the water. Sometimes gives subdued whistled calls. Fairly common on high lakes, used to be regular at Huacarpay lakes but now rare or non-existent there, perhaps due to introduction of exotic fish species. Still common on deep lakes in the Chinchero area such as lakes Piuray and Huaypo, often in large concentrations in winter.

FLAMINGOES Phoenicopteridae

Flamingoes are elegant wading birds with long legs and necks and a uniquely shaped bill. Most have a preference for brackish or saline water. They feed on microscopic organisms by filtering water through their specially adapted bills, with head held low and the bill upside down underwater. They nest in large colonies laying one egg in a conical clay mound.

Chilean Flamingo   *Phoenicopterus chilensis*
95-105 cm. > 3000 m. Identifying flamingoes is not an easy matter (three species being possible – Chilean, Andean and James's) especially with immature birds, and great care should be taken. This species is recognized in the adult plumage by reddish "knees" contrasting with gray legs and also pinkish (not yellow) base to bill. Immature carefully identified by lack of pink in the plumage and amount of black on the bill. This species is more likely to be seen in freshwater situations, being less tied to saline lakes and to date is the only species recorded in our area, notably at Huacarpay lakes south of Cusco, where it is rare.

 White-tufted Grebe
 juv

Silvery Grebe

Chilean Flamingo

STORKS   Ciconidae

Storks are large wading birds with a worldwide distribution in warmer areas. They are long-legged birds resembling herons but are much less dependent on water. They have heavier bills and, unlike herons, they fly with their necks fully extended. They soar to great heights and migrate long distances. They feed on invertebrates and small vertebrates, and build large stick nests in trees.

Jabiru   *Jabiru mycteria*
122-140 cm. Vagrant 2600-3000 m. A gigantic white stork with a very large black bill and head. Basal third of the neck is red. A bird of the marshy savannas and Amazonian rivers east of the Andes. May well migrate across the Andes based on the evidence of stray records. Alone or in quite large scattered groups. They stalk prey in grassy and marshy areas, sometimes at the edge of ponds while migrating. Has been recorded flying high and, on several occasions, resting at lakes in the Cusco area. There is one record of a Jabiru sitting on the Inca walls of Machu Picchu itself.

Wood Stork   *Mycteria americana*
89-101 cm Vagrant 3000 m. Readily recognized by large size, bare head and neck, white body with contrasting black remiges. Bill dark or yellow in the juvenile. Inhabits marshes and river margins. Only known from one record in 1986 at Huacarpay lakes south of Cusco. Vagrant from the Amazonian lowlands.

CORMORANTS   Phalacrocoracidae

Cormorants are foot propelled aquatic birds with long necks. They swim low in the water and dive for fish, remaining under water for considerable amounts of time. They are often seen conspicuously perched on prominent snags along shorelines with wings spread to dry, as unlike other water birds, they get soaked quickly. They fly well but have take off problems – running across the water surface with much flapping before getting airborne. They are colonial nesters.

Neotropic Cormorant   *Phalacrocorax olivaceus*
70 cm. 2000-4200 m. Bronzy black all over. Immature fuscous above and a dull pale brown below. The only cormorant in our region, therefore unmistakable. When fishing, can stay down for a long time and travel distances of up to 100 m under water. Can be found alone or in pairs or small groups on lakes (Piuray and Huacarpay lakes) and along rivers (Urubamba River). Seems to have benefitted from the introduction of exotic fish such as trout.

HERONS   Ardeidae

A well-known family, characterized by long necks, long legs and straight pointed bills. They are found near water for the most part, where they stand patiently or wade in the shallows. Many herons roost and nest communally and most stalk fish, amphibians and large aquatic insects. Nests are coarse platforms of sticks or reeds. Both sexes build the nest, incubate and feed the young.

Yellow-crowned Night-heron   *Nyctanassa violacea*
60 cm. Vagrant 3300 m. Shape as in previous species. Once a bird exclusive to mangroves in NW Peru, now steadily spreading south along the coast. One record of a long staying bird near Cusco. Gray with white cheek bordered by thick black stripes and yellow fore-crown. Immature brown above with whitish spots, below off white with brown vermiculations. Similar behavior to Black-crowned Night Heron.

Black-crowned Night-heron   *Nycticorax nycticorax*
62 cm. 3000-4700 m. Chunky with a short neck and heavy black bill. Adult: crown and back glossy black, long occipital white plumes. Immature: brown above, streaked and spotted all over. Inhabits marshy freshwater habitats with reeds, as well as overgrown drainage ditches and ponds. Usually found alone, sometimes in pairs. Crepuscular or partly nocturnal, but also encountered during the day in the Andes. Stands with a hunched posture at the edge of reed beds or in ditches waiting for prey. Seems to be a frog specialist. When flushed it flies long distances with measured wingbeats. Uncommon but can be encountered in a variety of moist localities, mostly at higher elevations.

Fasciated Tiger-Heron *Tigrisoma fasciatum*
65 cm. < 2200 m Vagrant to 3300 m. The *salmoni* subspecies is present. A dark, finely barred heron of fast flowing rivers. Juvenile: broadly barred black and tawny. Found on gravel bars and boulders along fast flowing sub-montane rivers and streams in humid areas. Usually found alone, occasionally in pairs, perched on rocks and boulders at the edge and in the middle of fast streams and rivers. Nervous and flushes easily. Often found in the shadows of overhanging vegetation. Uncommon at Machu Picchu but can be seen along the Urubamba River between Aguas Calientes and the Aobamba Valley.

Striated Heron *Butorides striatus*
40 cm. < 4000 m. Smallish heron. Found in all freshwater habitats but particularly where there is ample marsh and reed vegetation. Alone or in pairs. Stands on a perch just above the water or water's edge to hunt and very rarely wades. When alarmed it flicks its tail and raises its bushy crest. Regularly recorded at up to 4000 m although mostly a lowland species. Fairly common at Huacarpay lakes south of Cusco mostly November to March.

Cattle Egret *Bubulcus ibis*
50 cm. 3000-4000 m. Inhabits a variety of habitats and is very adaptable. Found in agricultural areas, especially where there are cattle, as well as humid habitats. Usually in groups, often feeding near livestock looking for large insects. Often hunched when at rest but extends neck when feeding. Tame. Fairly common and resident at Huacarpay lakes south of Cusco.

Cocoi Heron *Ardea cocoi*
110 cm. < 3500 m. A very large, lanky and slender-necked heron. White wrist patch in flight. Encountered along rivers, freshwater lakes and marshes. Usually solitary and wary. Stands and waits or wades in shallow water looking for fish and amphibians. Flight slow and labored. Rare in the Andes, one record, from Machu Picchu along the Urubamba River. A few scattered sightings at Huacarpay lakes south of Cusco.

Great Egret *Ardea alba*
110 cm. < 4000 m. This is a large, slender, long-necked, completely white heron of cosmopolitan distribution. Has a yellow bill and totally black legs. Found along rivers and near freshwater ponds and lakes. Solitary or widely-spaced individuals stand motionless in shallow water for long periods, waiting for frogs or fish. Rests with other egrets in mixed species groups. Recorded as a non-breeder at elevations of up to 4000 m but apparently breeds at Huacarpay lakes south of Cusco.

Snowy Egret *Egretta thula*
55-66 cm. 2600-4000 m. A smaller white heron with black bill and legs. The feet and lores are bright yellow. Inhabits freshwater marshes, ponds and rivers. Feeds actively in shallow water, walking and high stepping in search of fish, insects and amphibians. Usually encountered in small, loose groups but also found alone.

Little Blue Heron *Egretta caerulea*
56-66 cm. 3000 m. A medium sized Heron. Immature is all white with greenish legs. Intermediate plumaged birds can be seen. In all plumages, note relatively thick bicolored bill. Historically unrecorded in the region but with increasing spread of its range, this species has been a regular feature of Huacarpay Lakes since 2000 with evidence of breeding (Venero 2007).

IBIS   Threskiornithidae

A well known group of birds found worldwide. They mostly live on grassy and marshy plains and have long, slender decurved or spatulate bills. They are characterized by bare facial skin and, unlike herons, fly with outstretched necks. They use their long bills to probe for crustaceans and other small prey. Large stick nests are built in colonies and both sexes incubate and feed the young.

Puna Ibis   *Plegadis ridgwayi*
60 cm. 3000-4800 m. Long curved dark-red bill. Legs black. When breeding the plumage of the head and neck is a rich chestnut. Juveniles show some white streaking. Found in marshy areas and reed beds near lakes and rushy wet meadows and pasture, mostly where the ground is damp but sometimes on hill slopes a long way from water. Mostly in flocks, sometimes quite large, walking in loose association, with a hunchbacked appearance. Flies in tight groups or wavy lines often low over the ground. Named for Robert Ridgway (1850-1929) US ornithologist and curator of birds at the Smithsonian Institution.

Andean Ibis   *Theristicus branickii*
75 cm. > 3500 m. Sometimes thought to be a subspecies of Black-faced Ibis. Observations (Vizcarra 2009) suggest that the two taxa segregate where they occur sympatrically during nonbreeding. A thickset ibis with broad wings and a short tail. Inhabits open marshland and grassy plains, as well as puna grassland with bunchgrass. Can be found alone, in pairs, or in small family groups. Walks while feeding and probes into tussocks. Flies low over terrain and wanders long distances throughout the day to feeding sites. Quite noisy, especially when flying – makes a clanking metallic *'quank - quank.'* Uncommon at Machu Picchu but may be looked for at Pampacahuana near the Salcantay massif and regular near the pass at Abra Malaga and higher areas of the Sacred Valley.

AMERICAN VULTURES   Cathartidae

Vultures are large scavenging birds with naked heads. Their short weak bills are designed for eating carrion. Their chicken-like feet are not suited to carrying prey. Usually silent, they feed on carcasses and garbage. They soar with great proficiency and find most of their food by keen eyesight. They nest in inaccessible rocky terrain and the young take a long time to mature.

Turkey Vulture   *Cathartes aura*
75 cm, wingspan 180 cm. Bare skin of head reddish in adults, browner in immatures. Wings, two-toned from below with dark under-wing linings and pale-gray flight feathers. In flight tilts from side to side while gliding on wings held well above the horizontal. Mostly a lowland species inhabiting open country and larger river margins. Sometimes dives or glides with set wings. Can find prey by smell unlike other vultures. Recorded as a casual visitor at elevations of up to 4000 m, but mostly below 2500 m. A wanderer to our region and uncommon.

Andean Condor   *Vultur gryphus*
120 cm, wingspan 300 cm. Huge – weighs 11 kg. In flight recognized by long rectangular "fingered" wings and short tail. Adult: black with a silvery-white panel on the upper wing. Immature: all dusky brown with short brown down on the head, gradually showing paler panel on the upper wing as age advances. Inhabits desolate areas with high, steep mountains and cliffs. Encountered singly or in pairs, sometimes larger groups near carrion. Often soars for hours above hillsides on horizontal or very slightly upraised wings. "Fingers" often strongly upcurved. When gliding sometimes crooks wings. Perches on rocks high up on mountainsides. Found along the Inca Trail between Llactapata and Warmiwañusca Pass, but also can be seen from the Machu Picchu ruins as well as higher parts of the Sacred Valley including Abra Malaga.

Puna Ibis

Andean Ibis

Turkey Vulture

juv

Andean Condor

OSPREY Pandionidae

Found worldwide, ospreys are always found near water (except during migration) and feed exclusively on fish. A boreal migrant to South America (seen from October to April), but immatures seen at any season. *Pandion cristatus* Eastern Osprey is considered a distinct species from Western Osprey *(P. haliaetus)* by some authorities (Wink *et al.* 2004, Christidis & Boles 2008).

Osprey *Pandion haliaetus*
50-66 cm. 3000 m. From below looks white with large black patch on wrist and distinctive crook to the wing. Hovering heavily over water, they plunge-dive for large fish. When migrating found mostly along rivers. Known in our region from Huacarpay lakes south of Cusco where it is rare during the boreal winter. Named after the Greek King of Athens, Pandion, whose daughters were metamorphosed into a Nightingale and Swallow respectively.

HAWKS Accipitridae

A worldwide family of diurnal predatory birds characterized by strong hooked bills and gripping feet with long curved talons. Flight and tail feathers are often barred, but identification between species can be tricky. Hawks take live prey ranging from insects to fairly large mammals. Kites and harriers are more maneuverable and have specialized feeding habits. Hawks live at low densities and hold large territories. Females are larger than males and, in most species, the sexes have different roles – males hunting for smaller prey near the nest site to feed the young, while the female hunts over a larger area for bigger prey. Most build bulky stick nests and lay a small clutch of eggs. The young are covered in white or gray down and are fed for weeks to months at the nest and after fledging.

Swallow-tailed Kite *Elanoides forficatus*
60 cm (wingspan 130 cm). < 2600 m. vagrant to 4000 m. With its white body and black, deeply forked tail, this kite is unmistakable and one of our most striking birds of prey. Found in all humid forested regions, mostly in humid sub-montane forest. Wanders widely and is mostly aerial. Gliding gracefully above mountain ridges, it is often found in groups of up to 30 or more. It feeds on the wing, catching large insects and small vertebrates from the tree canopy or in the air. Its call is an infrequent *'knee-knee.'* Uncommon at Machu Picchu.

Black and Chestnut Eagle *Spizaetus isidori*
60-80 cm (wingspan 175 cm). 1800-3500 m. Very big, the female being the largest by far. The crest is often raised and the tarsi are feathered. In flight appears all dark with distinct light gray patches at the base of the primaries. Immature shows pale gray head and darker crest. Gray above and white below with streaking. Tail with three narrow dark bars and broad terminal bar. Inhabits humid premontane forest especially in large valleys. Seen soaring and gliding across valleys on horizontal wings. Sometimes perched on an exposed branch in the canopy. Call – *'chee-chee-chee.'* Feeds on arboreal mammals and large birds. Uncommon at Machu Picchu. Named for Isidore Geoffroy Saint-Hilaire, French zoologist (1805-1861).

Semicollared Hawk *Accipiter collaris*
30-36 cm. < 2500 m. Brownish black above, with a broken white collar on the hind-neck. Inhabits humid sub-montane forest. Call a series of descending *'kee'* notes. Rare in humid forest at Machu Picchu.

Plain-breasted Hawk *Accipiter ventralis*
26-35 cm (wingspan 55 cm male or 65 cm female). 1000-3500 m. A small hawk with a square tail. Thighs rufous. Inhabits humid forest borders, forest edge wood-lots and second growth. Perches, semi-concealed in foliage, then dashes through dense cover in pursuit of small birds. Occasionally glides and soars on horizontal wings. The call is a rapid *'kee-kee-kee'* indistinguishable from the preceding species. Most commonly seen below 3000 m. Closely related to the Sharp-shinned Hawk *(Accipiter striatus)* and, by some authorities, included in this superspecies.

Cinereous Harrier *Circus cinereus*
45-50 cm. (wingspan 120 cm). 2500-4500 m. Note the small head and long, narrow wings with four "fingers" held above the horizontal and the white rump in both sexes. Found in rushy fields, marshes and reed-beds and cultivated fields of wheat and barley. Flies low over grassy hill-slopes and open fields. Glides and soars with raised wings. Nests on the ground in rushes or grass and roosts communally on hillsides. Only harrier likely to be found in Peru.

Montane Solitary Eagle *Buteogallus solitarius*
66-70 cm. < 2200 m. Large with very long broad wings and very short tail. Under-wings are dark. Immature: heavily streaked with buff and brown, with some black patches on the thighs and chest. Inhabits humid, forested foothills in pre-montane regions. Alone or in pairs, soaring heavily on horizontal wings over forested hills or in long glides down steep, forested mountain valleys. Perches on high exposed branches. Takes snakes and small rodents. The call is a piercing *'peep-peeeep-peeeep'* and *'yeep-yeep-yeep.'* Rare at Machu Picchu. Best looked for in the Mandor and Aobamba Valleys.

Roadside Hawk *Rupornis magnirostris*
33-40 cm (wingspan 68-92 cm). < 3000 m. The *occidus* race is present in our region. A common hawk of the lowlands, and at the edge of its elevational range at Machu Picchu. A small yellow-eyed buteo. A slow-moving hawk, mostly seen perched on a twig or exposed snag. Flies weakly and seldom very far with rapid flapping and alternate glides. Soars infrequently with much flapping, often calling. Mostly feeds on snakes and other reptiles, and sometimes insects. The frequently given call is a descending *'sweeeeeeeeeeee.'* Mostly a lowland species but ascends to 3000 m in Peru. Can be seen along the Urubamba River near Aguas Calientes.

White-rumped Hawk *Parabuteo leucorrhous*
37 cm (wingspan 85 cm). 1800-3300 m. Adult: a short winged black buteo. Mostly black with a white rump and undertail coverts. From below, body black, flight feathers blackish, wing linings whitish. Immature: mostly brown mottled rufous above, creamy buff below, upper and under tail coverts white, tail barred rufous. Inhabits dense humid pre-montane forest. In flight circles low over forest, perches on low branches in the sub-canopy. The call is a short whistled scream. Can be seen in the Mandor Valley near Machu Picchu and other forested areas of the Sanctuary.

Variable Hawk *Geranoaetus polyosoma*
47-59 cm (wingspan 113-151 cm). 1800-4600 m. *G. polyosoma* (Red-backed Hawk) and *G. poecilochrous* (Puna Hawk) were often treated as separate species. Genetic data (Riesing *et al.* 2003) are consistent with hypothesis that *G. polyosoma* and *G. poecilochrous* are conspecific. However, Cabot & de Vries (2004) and Cabot *et al.* (in press) present additional data that support their recognition as separate species. Typical buteo shape. Very variable but all adults have a white tail with a black sub-terminal band. There are two color phases, dark and light and a confusing series of variations as a result of a complex moulting process from juvenile to immature, through to adult. Young birds are very variable but usually dark brown above, buff below streaked on the throat and breast and barred on the lower under-parts with brown. Found on open hill-slopes with some brush and bracken as well as partly cultivated slopes. Avoids dense forest. Often hovers with heavy, floppy wingbeats or hanging hover in the wind. Soars with flat wings and flies with shallow, stiff wingbeats. Hunts small rodents and insects. Call *'kyeea-kyeeah.'* Can be seen near the Machu Picchu ruins and throughout the Cusco and Sacred Valley area.

Black-chested Buzzard-Eagle *Geranoaetus melanoleucus*
65-80 cm (wingspan 175-200 cm). 2500-3800 m. A large chunky eagle-sized hawk. Female is much larger than the male. Adult: flight profile triangular due to its very broad based wings and a wedge-shaped tail. Juvenile: longer-tailed, but still with broad wings, uniformly brown above and tawny below with some mottling. Black primaries. Found in semi-arid and open country, inter-montane valleys and rocky open slopes. Sometimes, but rarely, spotted in more humid terrain. Found alone or in pairs perched on rocks or bare branches in isolated trees. More often seen soaring effortlessly over ridges on horizontal or slightly upturned wings. Hovers on occasion. Eats snakes, small rodents and sometimes other birds. Call is a drawn-out *'keeeuuu.'* Common along the Inca Trail between Llactapata and Wayllabamba, the Cusco area and the Sacred Valley of the Incas.

Broad-winged Hawk *Buteo platypterus*
34-45 cm (wingspan 80-100 cm). < 3000 m. A boreal migrant, present from October to March in small numbers. A small robust hawk. Tail has broad white bars. Underwing white, vaguely barred and with a distinctly dark trailing edge. Immature: below, white with dark streaks, flight feathers barred, five or six bars on the tail. Inhabits forested mountain slopes, second growth and partially-cleared areas. On migration it forms large "kettles" in thermals. It soars and glides in thermals with horizontal or slightly-arched wings. Hunts by waiting on a low perch at the forest edge. Rarely seen at Machu Picchu.

White-throated Hawk *Buteo albigula*
38-48 cm (wingspan 95 cm). 1800-3500 m. A short tailed and broad winged buteo. Adult: blackish brown above and on sides of head to below eyes giving a hooded effect. Immature: similar to the adult but with large dark brown spots on the breast and flanks. Found in forest adjoining open areas and clearings, elfin forest and scattered patches of semi-humid forest in ravines. Circles low over forested areas or glides across hillsides and valleys close to the treetops. Perches on exposed branches in the canopy. Call is a descending drawn out *'kee-aaa.'* Can be seen below Phuyupatamarca ruins along the Inca Trail and at Abra Malaga, also casually in the highlands near Cusco.

FALCONS   Falconidae (including Mountain Caracara)

The falcons are a diverse group of birds of prey with a cosmopolitan distribution. They differ from hawks and eagles in that, among other things, they have a "tooth" or notch on the upper mandible. Ericson *et al.* (2006) and Hackett *et al.* (2008) found that the *Falconiformes* are actually more closely related to the *Psittaciformes* and *Passeriformes* than to any other orders. The caracaras are strictly New World and have long wings and tails. Omnivorous and opportunistic, they occupy the role of crows in South America. True falcons are streamlined, with tapered wings, predators capable of fast flight and tremendous speed. They have well developed flight displays.

Mountain Caracara *Phalcoboenus megalopterus*
44-19 cm. wingspan 111-124 cm. 3200-4700 m. Immature: completely dark to light brown with pale buff rump. Broad white buff zone across the base of the primaries, conspicuous in flight. Found in open puna grassland and at higher elevations. Feeds on the ground in open plains, grazed areas and small fields. Often in pairs or small groups walking on the ground, especially in plowed fields, looking for insects. In flight wingbeats are shallow and stiff. Sometimes many birds congregate in large flocks. Nests on rock ledges. Fairly common in our area in non-forested places.

American Kestrel *Falco sparverius*
25 cm. wingspan 55 cm. The *cinnamominus* subspecies is present at Machu Picchu. A small well known falcon. Found in a variety of habitats – forest edge, dry areas with xerophytic plants, open country with scattered trees or rocky outcrops. Often in villages. Avoids densely forested areas. Usually encountered alone or in pairs. Fast flight with stiff wingbeats, alternating with flying and soaring. Hovers frequently, then stoops on mostly insect prey. Often seen perched conspicuously on wires, posts or rocks. The call is a high then *'klee-klee-klee.'* Common.

Orange-breasted Falcon   *Falco deiroleucus*
33-38 cm. wingspan 69-85 cm. < 2000 m. A largish falcon with long pointed wings. Unlike the two preceding species, found in forested country often in clearings or along rivers. Single birds or pairs spend a lot of time perched on dead branches. Flies fast and rapidly when hunting birds taking them on the wing. A lowland falcon, uncommon, but regular at Machu Picchu, where it breeds.

Aplomado Falcon   *Falco femoralis*
37-42 cm. wingspan 76-102 cm. 2400-4300 m. Long tailed and streamlined. Tail projects beyond wingtips at rest. Inhabits open high country but also semiarid inter-montane valleys and *Eucalyptus* groves. Usually encountered in pairs, occasionally alone. Often hunts in pairs, one bird flushing prey and the other cutting off the escape route. Flies very fast and with agile turns. Perches on rocks, posts and other exposed features. Calls include a high pitched *'cree-cree-cree'* also *'ree-ree-ree.'* Fairly common.

Peregrine Falcon   *Falco peregrinus*
38-42 cm, wingspan 95-117 cm. > 1800 m. Female significantly larger than male. Two races can occur in our region. Between October and April, the arctic boreal migrant *tundrius* and throughout the year, the resident race *cassini*. A powerful falcon with a shorter tail than that of the Aplomado Falcon, not extending beyond the wingtips when at rest. In all plumages with a dark hood covering almost the entire head (except chin and throat, in *cassini*) or well below the eye with a black mustache mark contrasting with white throat and cheeks *(tundrius)*. In flight wings pointed, underwing densely barred and spotted. Found in open rocky terrain in the Machu Picchu area. Alone or in pairs. Takes most of its prey in the air. Perches conspicuously in trees or on rocks. Call a high pitched *'yeek-yeek-yeek.'* At all elevations.

RAILS   Rallidae

Most members of the "typical" rail family are little known, secretive residents of marshes, swamps and tall wet grasslands. Some are also crepuscular and are, in general, rarely detected except by call. With compressed bodies, they are well suited to slipping through vegetation. Their diet is varied but mostly invertebrates. Unlike rails, gallinules and coots are aquatic and more easily seen. They are excellent swimmers and prefer a diet of aquatic vegetation. Nests are well concealed cups of reeds and sedges.

Ocellated Crake   *Micropygia schomburgkii*
14 cm. Vagrant. 2100 m. A very small rail. Included here as a vagrant on the basis of one trapped near Wiñay Wayna ruins in 1996 on a scrubby, bushy hill slope. Normally inhabits open savannas and grassy plains. Terrestrial. Call is a long series of *'pr-pr-pr'* notes.

Paint-billed Crake   *Mustelirallus erythrops*
18 cm. 3000-3400 m. Small and gray-breasted, with prominent red and yellow bill. Calls include churring and *'pip'* and *'aark'* notes. Inhabits lake margins and wet fields where it is secretive, creeping around near the ground. At dawn, it will sometimes come into more open muddy areas or floating grass. In our region, known from Huacarpay Lakes and the Pampas de Anta. Rare and almost certainly not resident.

Spotted Rail   *Pardirallus maculatus*
26 cm. Vagrant. 3000 m. One record at Huacarpay Lakes. Bold black and white plumage in adult. Immature much browner and duller with fewer markings. Habitat and behavior in our region similar to following species. Song a drumming like *'pum-pup-pum-pupppppppum.'* Also high pitched chatters and squeals.

Plumbeous Rail   *Pardirallus sanguinolentus*
35 cm. < 4000 m. The *tschudi* race is present in our region. A long-billed dark rail. Found in reed beds and grassy marshes, sometimes quite small ones. Also found along overgrown drainage ditches in cultivated areas. Usually alone or in pairs at the edge of reed beds, but often conspicuously out in the open feeding with characteristically cocked tail. Less skulking than most rails. Common at Huacarpay Lakes.

Purple Gallinule *Porphyrio martinica*
28-32 cm. < 3200 m. Immature: Dull. Brownish above, with bluish tinge on wings and buffy white below. Legs dull yellow. Usually seen walking on floating vegetation of lakes and marshes with cocked tail. Climbs around in lakeside vegetation. Has various cackling and clucking calls. Regularly recorded at Huacarpay Lakes, but seemingly less common in recent years.

Common Gallinule *Gallinula galeata*
36-38 cm. 2200-4400 m. The large highland race *garmani* is present in our region. Immature: dark gray, paler below, becoming whitish on the throat and belly. Bill dusky flesh colored. Inhabits ponds, marshes and lakes with plenty of marsh vegetation and submerged or floating weeds. Also on muddy lake shores. Feeds on open water pecking at the surface. Hides in dense cover when disturbed. Walks and swims with a jerky motion, flicking tail. Makes a variety of clucking and chattering calls and louder trumpet-like calls when disturbed.

Andean Coot *Fulica ardesiaca*
43 cm. 2500-4600 m. Stockier than Common Gallinule. Frontal shield from deep maroon (commonest in the Cusco area) to orange yellow or white. Immature is dark gray, paler below with a mainly white face. Found on lakes and marshes, generally on larger bodies of water than the preceding species. Mostly found where there is some reed and marsh vegetation but also on barren lakes. Gregarious, often found in very large flocks in suitable habitat, foraging in shallow water with dense submerged or floating vegetation. Call is a low *'rrrr'* or *'lurp.'*

Giant Coot *Fulica gigantea*
48-64 cm. 3900-4600 m. A massive coot, dark slaty-colored all over with red feet Immature: drab gray with dull red legs and feet. Usually found in pairs or families. Immature can disperse by flight, but adults are too heavy to fly. Adults may walk from one lake to another. They build huge nests on barren ponds and lakes with water weed that protrude half a meter from the surface of the water, often used for several generations. Male makes a variety of gobbling and growling sounds. Females: a gentler *'chee-jeerh'* or *'hi–hirr–hiirr.'* Fairly common on high altitude lakes in the Cordillera Urubamba above the Sacred Valley of the Incas, for example Laguna Muchillay above Lamay.

SUNBITTERNS   Eurypygidae

A strange rail or heron like bird of shady streams and river margins of eastern Peru with un-webbed toes. The drab colors belie a "sunburst" of color on the flight feathers, conspicuous in flight and during the spectacular display with wings and tail fanned. Nest a globular mass of vegetation, moss and mud usually a few feet up on a bush at water's edge.

Sunbittern *Eurypyga helia*
43 -48 cm. < 2200 m. Rare with only one record along the Urubamba river near Machu Picchu. Other sightings of the *meridionalis* foothill race could be expected. Swims low in the water. Singles or pairs inhabit the edge of streams and rivers, often on fast moving rock-lined streams where they walk quietly along the shore often lunging for small animal prey. Flies with characteristic shallow wingbeats.

PLOVERS   Charadriidae

Plovers are compactly built, small to medium sized shorebirds. Plumage can be boldly patterned but some have reduced patterns and a basic (non-breeding) plumage. Less sociable than other shorebirds, often found alone or in pairs. Most species live in small flocks outside the breeding season. Nest is a shallow scrape. Young are precocial and extremely well camouflaged.

Andean Lapwing   *Vanellus resplendens*
33 cm. 3000-4600 m. A large, well known plover and a conspicuous element of high Andean bird communities. In flight note broad rounded wings and distal half of tail black, large wing patch and basal half of tail white. Found on open parts of marshes, rushes, wet pasture and boggy ground, sometimes along lakeshores and on short grassy plains. Very noisy – individuals and pairs mob intruders. Wingbeats are rapid and shallow. On alighting stands for some time with wings raised. Feeds mostly in pairs walking and picking from the ground surface. Very vocal especially when disturbed. – *'cree-cree-creee'* or a gull-like *'ka-leek ka-leek ka-leek,'* etc. Fairly common, forms small winter flocks.

American Golden Plover   *Pluvialis dominica*
24-27 cm 3000-4500 m. The *Pluvialis* plovers are not members of the traditional plover family *Charadriidae* (Baker *et al.* 2007) and may deserve their own family. Boreal migrant. Present mostly from August-November and March-April. During March-April migration, black feathers emerge on all of underparts. In flight buffy-gray wing linings and only faint wing bar. Call single or two note whistles *'ki-eee'* or *'chee-it.'* Habitat in the Andes is grassy meadows and plains near lakes. At Huacarpay Lakes flocks of up to 50 have been recorded in September.

Black-bellied (Grey) Plover   *Pluvialis squatarola*
27-30 cm. 3000 m. Boreal migrant, mainly September to April. In basic plumage resembles the previous species but chunkier and heavier billed, paler. In flight shows conspicuous white wing-bar and white underwing contrasting with conspicuous black auxiliaries. Call a plaintive *'plee-oo-eeee.'* Vagrant and only recorded in our region at Huacarpay Lakes south of Cusco so far. May be more common than thought on northward migration, when black feathers show through on breast, neck and face.

Semipalmated Plover   *Charadrius semipalmatus*
17-19 cm. 3000-4100 m. Boreal migrant, mostly noted in our region September to December in basic non-breeding plumage. Note stubby black bill with yellow base and yellow legs. In flight shows long white wing stripe. Calls – a piping *'CHEE-weet'* or *'keet.'* Uncommon on flat areas near lakes during the austral summer but also recorded on small ponds at high altitude during migration.

Puna Plover   *Charadrius alticola*
17 cm. 3300-4500 m. A compact, "neckless" plover. Legs black. Non-breeding birds lose part of the black pattern. In flight shows half a wing stripe on the inner primaries and white sides to the tail. Usually encountered on wide expanses of firm mud or dried out lake beds and heavily grazed shorelines. Can be seen alone or in small loose flocks. Runs very fast or flies low over the terrain when threatened. Calls include *'tseet'* and *'prit.'* Recorded at Huacarpay and Piuray Lakes as a vagrant, but should be looked for on higher lakes throughout the region.

Collared Plover   *Charadrius collaris*
14-16 cm. 3000-4500 m. Similar to the previous species. A rare visitor to highland lakes and marshes. Trans-Andean migrant? Calls a clear *'peet.'* This Amazonian species has been recorded in our area in October and November on lake shores and mudflats around lakes.

Diademed Sandpiper-Plover   *Phegornis mitchellii*
17-19 cm. > 4000 m. Unmistakable. Bill long and thin more like Sandpipers than a Plovers. Despite striking plumage can be difficult to see with its cryptic plumage and quiet habits. Rare in our region with records from Abra Malaga. With lots of suitable habitat it could be expected though it is notorious for being absent in what is apparently good habitat – mossy peat bogs especially with running streams and cushion plants. Named for David William Mitchell (1813-1859) English zoologist.

STILTS   Recurvirostridae

A worldwide family related to plovers. They have small heads, long necks, long thin bills and very long legs. They nest in loose colonies on mud or tussocks in shallow water. The young develop quickly and find their own food and run and swim at an early age.

White-backed Stilt   *Himantopus melanurus*
36-41 cm. 2500-4200 m. The form occurring in the Cusco and Machu Picchu area is sometimes considered conspecific with Black-necked Stilt *(Himantopus mexicanus)*, as intermediate plumages occur. Very lanky with a needle-thin straight bill and extremely long, slender pink legs that trail behind tail for 15 cm in flight. Inhabits lakes, ponds and open marshlands. Gregarious. Strides, gracefully lifting feet high in shallow water, sometimes swimming. Feeds by pecking and snatching in soft mud. Sometimes sits on the ground. Very vocal giving a series of high pitched *'yip'* notes with variations. Mostly occurs between October and May, not proven to breed in our region.

Andean Avocet   *Recurvirostra andina*
43-46 cm. > 3000 m. Strongly upcurved bill. Mostly white with back, wings and tail black. Call a sharp barking *'kluu-kluut.'* Usually an inhabitant of saline lakes, this species has been recorded as a rare vagrant on Huacarpay Lakes south of Cusco, but post breeding wanderers should be looked for on any high altitude wetland in the region.

SANDPIPERS AND SNIPES   Scolopacidae

Most members of this family breed in the Northern Hemisphere but winter in South America. Most sandpipers are slim with longish necks and bills. They probe soft ground for invertebrates and mollusks. In South America, usually found in winter plumage and in flocks. Only snipes breed in the Andes and are notable for their aerial display, where the modified flight feathers are used for "drumming." Phalaropes are dainty sandpiper-like birds found at sea outside the breeding season. They swim easily.

Puna Snipe   *Gallinago andina*
23-25 cm (incl. 5-6 cm bill). 3000-4500 m. Intricately speckled and vermiculated buff and black. Tail rufous with white sides barred with black. All flight feathers have white tips. Legs yellow and short and do not trail behind the tail in flight. Inhabits boggy parts of puna grassland, marshes, wet meadows and muddy ditches. When flushed rises steeply and flies fast, quickly dropping into cover. When flushed makes a *'sketch'* call. Call delivered while on the ground is a *'dyak dyak dyak'* or *'dyuc dyuc dyuc.'* Displays in wide circles with shallow dives while "drumming." Fairly common where there are reeds and marshy ground.

Andean Snipe   *Gallinago jamesoni*
30 cm (incl. long 8-9 cm bill). 3000-3800 m. A heavy bodied snipe. No rufous or white in the tail. Found in boggy grassland at tree line with islands of bushes and trees and a general mosaic of bogs, bamboo, elfin forest and grassland. Crepuscular and nocturnal. Stays hidden in the vegetation by day and when flushed quickly hides in cover, often making a *'tzic'* call. Displays just after dark or just before sunrise. The display consists of flying slowly in wide circles with shallow wingbeats uttering a loud *'witchew-witchew-witchew-witchew'* for long periods, interspersed with 2-3 second "drumming" dives (sounds like a distant jet plane). Also calls from the ground – *'djic-djic-dyic'* for long periods. Quite common at tree line along the Inca Trail (Sayacmarca ruins) and at tree line at Abra Malaga.

Imperial Snipe   *Gallinago imperialis*
30 cm (incl. 9 cm bill). 2800-3500 m. A hefty, dark rufescent snipe boldly barred with black throughout. In flight note very broad rounded wings and very short tail. Found on mountain ridges at tree line with wet elfin forest, sphagnum bogs and bamboo fringed glades. Crepuscular and nocturnal. Flies in wide circles pre-dawn and after dusk and well into the night giving a 10-second series of rough staccato notes that rapidly increase then decrease in volume. Then makes a short dive, producing a clearly audible rush of air. Displays from July to September at least. Less common than the preceding species. Should be listened for when camping in any suitable habitat. Occurs at Abra Malaga and along the Inca Trail to Machu Picchu.

Hudsonian Godwit *Limosa haemastica*
38-41 cm (incl. 7-9.5 cm bill). 3000-3500 m. Boreal migrant September to April. Long slightly upturned bill. Breeding plumage adopted during migration, chestnut below with barring. In flights shows a long white wing bar and white tail with broad black band at the tip. Very rare in our region with records September-November at Huacarpay and Piuray Lakes near Cusco.

Whimbrel *Numenius phaeopus*
40-45 cm (incl. 6.5-9.5 cm bill) Boreal migrant. Head striped and crown dusky with buff mid-stripe and buffy supercillium contrasting with dark eye line. Long decurved bill. Rare vagrant recorded at Huacarpay Lakes.

Upland Sandpiper *Bartramia longicauda*
18-20 cm (incl. 3 cm bill). 3000-3500 m. Boreal migrant recorded September to November and February to April. Note "pigeon-like" head with big eyes, slender neck and quite long tail. Calls *'kip-kip-kip ip-ip'* Found on short grassy areas such as soccer fields and the like. Has been recorded regularly at Huacarpay Lakes south of Cusco, mainly in September.

Spotted Sandpiper *Actitis macularia*
18-20 cm. > 1800 m. Boreal migrant from North America present mostly from August to March, but some birds stay year around. Breeding plumage: similar but thickly spotted below with large black round spots. In flight note white wingstripe and distinctive way of flying with stiff shallow wingbeats interspersed with short glides and wings bowed downwards. On the ground teeters persistently as though not well balanced. Found along rivers, streams and ponds. Usually encountered alone, inconspicuously feeding among rocks and debris at the water's edge. Call is a high pitched *'peet'* or *'peet-weet-weet.'* Mostly at elevations of up to 3300 m but can be seen higher.

Greater Yellowlegs *Tringa melanoleuca*
30-36 cm (incl. 6 cm bill). 1800-4500 m. Boreal migrant from North America present mostly from August to March, but some may remain throughout the year. Almost identical to the following species in plumage but larger and plumper with a heavier bill which is longer and slightly upturned (longer than the length of the head). Found along rivers, lakes and marshes, where it can be seen feeding actively in flocks or alone, sometimes associated with other shorebirds. Wades in shallow water and sometimes swims. The call is a distinct 3-4 note *'tew-tew-tew.'* Can be encountered at all elevations.

Lesser Yellowlegs *Tringa flavipes*
25-28 cm (incl. 3.5 cm bill). 1800-4500 m. Boreal migrant from North America present mostly from August to March, but some may remain throughout the year. Slender and straight black bill. In flight shows dark wings and whitish rump and tail. Found along rivers and margins of lakes and marshes. Wades actively and sometimes associates with other shorebirds. Less confiding than Greater Yellowlegs. The call is a one or two note *'tew-tew'* – somewhat less forceful than the Greater Yellowlegs. At all elevations.

Solitary Sandpiper *Tringa solitaria*
19 cm (incl. 3 cm bill). < 4000 m. Boreal migrant from North America present mostly from August to March. Slender black bill and long dusky-green legs. Obvious white eye-ring. In flight note lack of wing stripe, dark rump and barred sides to the tail. Inhabits all kinds of freshwater situations, especially where overhanging trees are present. Usually, as the name suggests, alone, but scattered groups of 3-4 may be found in favorable areas at the right time of year. Never in flocks as such. Nods its head continually as it wades in shallow water. Flight is erratic and fast. Call is a high pitched *'peet'* or *'peet-weet-weet.'*

Sanderling   *Calidris alba*
19 cm (including 2.5 cm bill). 3000-3500 m. Boreal migrant expected September to April. In breeding plumage which may be evident on northern migration, rusty and rufous tones on the upperparts, neck and upper breast. Call is a dry *'kit.'* Rare in our region but several records from Huacarpay Lakes south of Cusco especially in September.

Least Sandpiper   *Caldiris minutilla*
13-15 cm. 3000-3500 m. Boreal migrant casual in the Andes September to April. Small. Thin needle-like bill with slight droop. Call a drawn out *'three-eeet'* or *'priiit.'* Likes muddy lake edges and short wet grassy areas. Uncommon.

Baird's Sandpiper   *Calidris bairdii*
18 cm. 3000-4500 m. Boreal migrant from North America present mostly from August to March. At rest the long wings extend slightly beyond the tail unlike others of the genus. Found by lakes and marshes on the puna, dry lake shores and wet meadows as well as boggy valley bottoms. Usually in small flocks feeding mostly on land by pecking rather than probing. Call is low and raspy – *'kreet'* or *'kreer.'*

White-rumped Sandpiper   *Calidris fuscicollis*
16 cm. 3000-3500 m. Boreal migrant. Thick slightly drooping bill same length as head. In flight thin, short white wing-stripe and white rump contrasting with dark tail. On northward migration may show signs of breeding plumage with black back feathers edged with chestnut and dark stripes and spots on breast and flanks. Call a mouse-like *'tzeeet'*, also *'pip-pip.'* Very rare visitor to our region.

Semipalmated Sandpiper   *Calidris pusilla*
13-15 cm. 3000-3500 m. A rare boreal migrant to our region. During northern migration may show some black and rufous scalloping above. Call a rough *'djirt'* or *'whee-whee-whee-whee.'* Has been recorded at Huacarpay Lakes south of Cusco.

Pectoral Sandpiper   *Calidris melanotos*
20-23 cm. 3000-3500 m. Migrant from North America present mostly from August to March. A largish long necked *Calidris*. In flight little or no wing stripe noticeable, rump whitish on sides with a dark center. Encountered in open areas of marshes and wet meadows. Also, on open grassland. Usually not in flocks but sometimes in loose aggregations. Rather snipe-like zig-zag flight when flushed. Makes a low pitched *'chreek'*, *'trrik'* and *'cir-eep.'* Mostly at 3500-4500 m. Uncommon.

Stilt Sandpiper   *Calidris himantopus*
21 cm (including 4 cm long bill). 3000-3500 m. Boreal migrant. On northern migration may show rusty ear coverts, black scalloped mantle and under-parts barred with dusky. Calls are single *'tew'* notes similar to Lesser Yellowlegs. Uncommon in our region.

Buff-breasted Sandpiper   *Tryngites subruficollis*
20 cm. 3000-3500 m. Boreal migrant. Small head and small thin bill. Feeds with a plover-like action. Calls include harsh *'prik'* and *'tric'* notes. Usually tame and found in short grassy areas at lake edges and fields including soccer fields and airports. Has been recorded at Huacarpay Lakes south of Cusco.

Wilson's Phalarope   *Phalaropus tricolor*
23 cm. > 2500 m. Boreal migrant from North America present mostly from August to March. Note the black needle-like bill. In flight has solid gray wings with no contrasting stripe, but a white rump and tail. Found on shallow parts of lakes and ponds. Floats buoyantly on the water and is very gregarious. Stirs up small insect prey by spinning, then picks insects off the surface. Can occur at all elevations but is more common on high elevation lakes above 3300 m. Fairly common during migration.

SEEDSNIPE   Thinocoridae

A somewhat peculiar family, the members of which remarkably resemble the sandgrouse of the Old World. Like sandgrouse, they have chicken-like bills, a plump shape, short legs and vermiculated plumage. In flight they zig-zag in snipe fashion and can be confused with sandpipers to which they are probably most closely related. They feed off vegetation and buds rather than seeds. In the breeding season they occur in pairs or small family groups, at other times in flocks. The nest is a scrape on the ground. Four eggs are laid and incubated by the female. Seedsnipes feign injury when a nest is approached.

Rufous-bellied Seedsnipe   *Attagis gayi*
29 cm. 4000-5000 m. In flight, shows no real contrast or pattern. Inhabits rocky slopes, scree and bleak alpine terrain but feeds on nearby cushion bogs and valley bottoms. In pairs and small to large groups (up to 80 in winter). Usually tame and confiding, walks with an upright stance. When flushed, flies fast in a zig-zag course for long distances. Call given in flight is a melodic *'gly-gly-gly'* or *'cul-cul-cul.'* Often found right at snow line. Within the Sanctuary can be seen around the Salcantay massif.

Gray-breasted Seedsnipe   *Thinocorus orbignyianus*
23 cm. 3500-5000 m. Inhabits puna grassland with bunchgrass, also rocky and stony places with cushion plants and herbaceous vegetation. In the breeding season, disperses in pairs and family groups. Has a dawn display flight. Sings from a hummock or rock, a musical *'pukleeoo pukleeoo pukleeoo,'* etc. Within the Sanctuary can be seen near the Salcantay and Wakay Willka massifs, common at Abra Malaga.

GULLS AND TERNS   Laridae

Gulls are familiar lake and shore birds with worldwide distribution. Feet are webbed. Gulls often soar at great altitudes and also float on water. They are scavengers taking insects and floating items. They nest in colonies on islands or marshy places. They mob intruders near their nests. The 2-3 eggs and downy young are cryptically colored. The young are fed by both parents. Terns do not scavenge and are thinner billed with longer tails and are more nimble and dainty.

Andean Gull   *Chroicocephalus serranus*
46 cm. 2000-4500 m. In the breeding season, has a black hood, outside the breeding season only a dusky ear spot and crescent near the eye. Nests on high Andean plains, lakes and bogs. Outside the breeding season, they also feed along rivers and on open grassland and agricultural fields. Quite social, but not in large flocks. Feeds on insects over grassland. Nests in dispersed colonies in open parts of reed marshes or on islands in old coot nests or builds floating nest among floating weeds. Can be seen along the Urubamba River and lakes at higher elevations. Breeds at Huacarpay Lakes. Common.

Laughing Gull   *Leucophaeus atricilla*
39-45 cm. Vagrant. 3800 m. Rare boreal migrant to the Andes. Wing tips lack white, wings are very long, projecting well beyond tail. Similar species all have more striking wing patterns with white primary flashes. Name derives from its maniacal, rising and falling, laughing call *'ha-ha-hahaha-how-haow-haaaw.'* Recorded at Lakes Huaypo and Piuray. May be overlooked.

Franklin's Gull   *Leucophaeus pipixcan*
35-39 cm. 3800-3500 m. Boreal migrant, in huge numbers along the coast and wanderer to the Andes. Possible October to May. Alternate plumage is with full black hood and bright red bill. Calls are yelps and piping calls and a high pitched *'ha-ha-ha ha-yip.'* Might be found in the Andes along lakes and rivers. Rare with only a few records, notably the Urubamba River and Huacarpay Lakes south of Cusco.

Large-billed Tern   *Phaetusa simplex*
37-40 cm. Vagrant. 3000 m. A subadult was present at Huacarpay Lakes October 2 – November 26, 2005. This large Amazonian tern has an unmistakable wing pattern and large heavy yellow bill and legs. Calls a gull-like *'keeeyah.'*

Black Tern   *Childonias niger*
22-25 cm. Vagrant. 3000 m. Boreal migrant October to April. Can be expected in basic plumage but has been recorded at Huacarpay Lakes in May in alternate plumage. Flight buoyant and erratic with frequent dips into water. Very rare in our region.

Common Tern   *Sterna hirundo*
33-38 cm. Vagrant . 3000 m. Boreal migrant. One record from Huacarpay Lakes of a basic plumaged bird (photographed). Typical tern – very buoyant direct flight. Wing pattern is the best field mark with dark secondaries and outer primaries forming a translucent triangle on inner primaries.

SKIMMERS   Rhynochopidae

Skimmers are tern-like, large and long winged. There are three species (Americas, Asia and Africa). They have large strange bills. Both mandibles are depressed laterally and knifelike with the lower mandible longer than the upper. They plough open water with bill open to catch fish and crustaceans. When not feeding they rest on sand bars or lake edges.

Black Skimmer   *Rynchops niger*
41- 48 cm. 3000 -4000 m. Breeds in the Amazon but found on the coast where it does not breed. A rare trans-Andean visitor to high altitude lakes near Cusco. Found alone or in small groups often with gulls. Bill bright red tipped black. Plumage black and white. Flies buoyantly.

PIGEONS AND DOVES   Columbidae

A worldwide family ranging from small to very large birds. They are both terrestrial and arboreal and occupy almost all habitats. They fly fast. The head is small and bill weak. They are generally of subdued hues, sometimes with a metallic gloss on the head, neck and back. Nests are frail stick platforms in trees or bushes. The young are fed by regurgitation. Their cooing calls differ in rhythm and are useful for identification.

Rock Pigeon   *Columba livia*
32 cm. < 2000 m. Feral version of the Old World Rock Dove. Many colors were developed over centuries of near-domestication. Those resembling their wild ancestors have the head and neck darker than the back, black bars on the inner wing and a white rump with a black band at the end of the tail. There are many other varieties with colors from pure white to rufous or blackish. Lives around towns and settlements. Feeds during the day in fields and open areas. Social.

Spot-winged Pigeon   *Patageioenas maculosa*
32 cm. 2000-4000 m. A large, broad-winged pigeon. On the wing, particularly conspicuous in flight, is a broad white band slanting backwards from the wing bend, contrasting with black carpal area. Tail dark gray with blacker distal areas. Found in semi-arid habitats such as open woodland and scrub, agricultural areas and small settlements with *Eucalyptus* trees. Does not like humid forest. Usually encountered in breeding pairs or small groups. Feeds on the ground and flies with shallow wingbeats. Circles with slow wingbeats during display flight. Primary song delivered form a tree top is a few low soft *'coos'* followed by a louder *'cooouh-cuh-coooh,'* first note low pitched. Appears to be increasing dramatically in numbers. Can be seen throughout the Sacred Valley and Cusco areas even in stands of exotic trees in the city.

Band-tailed Pigeon   *Patageioenas fasciata*
35 cm. 2000-3000 m. A large, heavy pigeon. Inhabits humid montane and pre-montane forest and ravines with forest. Very social, usually in small to large flocks, sometimes in pairs. Always in the canopy feeding on berries, leaves and blossoms. Large groups fly swiftly across the treetops sometimes very high. When flushed, wing claps. The song is a weak owl-like *'hoo-hooo-hoo.'* Can be seen at various points along the Inca Trail and at Abra Malaga.

Plumbeous Pigeon   *Patageioenas plumbea*
33 cm. < 2000 m. Inhabits humid sub-montane forest, forest borders and secondary growth. Usually encountered alone or in pairs, but will flock at fruiting trees. Mostly keeps hidden in the canopy, where it can be difficult to detect. The song is a distinctive *'hoo-coo-cu-cuuuuu.'* Also, a loud purring common to the genus. A lowland species that ranges into the pre-montane zone, only likely to be found in the Machu Picchu Sanctuary at the lowest elevations. Rare.

Eared Dove   *Zenaida auriculata*
23-28 cm. 2500-4000. The white tail tipped *hypoleuca* race is present. In flight shows graduated shortish tail with broad white tips contrasting with black median bar. Inhabits semi-arid open land and agricultural areas where there are groves of trees and bushes, hedgerows etc. Gregarious: roosts and nests communally in trees. Feeds on the ground. Flight very fast and more or less direct. Song is a low descending *'oo-whoo'* sometimes repeated several times. Common in non-forested and urban areas. Increasing at a phenomenal rate in our region.

White-tipped Dove   *Leptotila verreauxi*
29 cm. < 2500 m. The *decolor* subspecies is present in our region. Outer tail feathers blackish and slightly graduated, all but the central pair with broad white tips. Loral line and orbital skin red. In flight shows chestnut-rufous underwing. Found in more open woodland, advanced secondary growth and open bushy woodland with taller trees. Avoids arid areas. Usually encountered alone or in pairs feeding on seeds and berries on the ground, walking with a mechanical gait. Quite shy. When disturbed, walks away or flushes onto a low perch with audible whirring of wings, where it bobs its head and walks up the branch into deeper cover. The song is drawn out *'whoo-woooo'* with the quality like blowing across the top of a bottle. Can be seen near the town of Aguas Calientes and on the grounds of the El Pueblo Hotel at Machu Picchu.

White-throated Quail-Dove   *Zentrygon frenata*
34 cm. 1800-3000 m. A very large quail-dove. The nominate *frenata* race is present at Machu Picchu. Inhabits humid montane and pre-montane forest, undergrowth and well developed secondary growth. Usually encountered alone, sometimes in pairs. Mostly terrestrial, feeding on seeds and berries on the ground. When disturbed, flushes to a low perch or walks calmly away. Shy. The far carrying song is a very low pitched *'whooooo'* delivered at regular well spaced intervals from a bush or small tree. Skulking and difficult to see except early in the morning where it may be seen walking on trails, especially in the grounds of the Machu Picchu Pueblo Hotel where it comes to feeders.

Ruddy Quail-Dove   *Geotrygon montana*
22cm < 2000 m. An Amazonian quail-dove only recorded once so far at Machu Picchu. Reddish brown with stripes on the side of the face. Found in forest interior quietly walking along trails and feeding on the ground. When flushed makes audible wing noises and flies to a low perch. Sometimes walks away calmly. Mostly terrestrial, feeding on seeds and berries on the ground. Song is even pitched slow *'wooOOOooo'* fading at the end.

Maroon-chested Ground-Dove   *Claravis mondetoura*
22 cm. 1800-3000 m. Found in the wet understory and along the edge of humid montane and submontane forest where there is a predominance of flowering bamboo (*Chusquea* sp.) to which this species is strongly tied. Found alone or in pairs. Difficult to see, keeping in dense undergrowth. Nomadic and unpredictable, following the flowering cycle of extensive bamboo patches. Calls from deep cover – a deep *'wroop-wroop-wroop'* repeated for long periods. Not uncommon in flowering bamboo patches near the Machu Picchu ruins complex and at Abra Malaga.

Bare-faced Ground-Dove   *Metriopelia cecilae*
17 cm. 2500-4500 m. A small, compact dove. The *gymnops* race is present in our region. Inhabits arid and semi-arid regions, open rocky and sandy areas with some herbaceous plants or cactus. Also, lightly wooded slopes and around human habitation where there is no plant cover. Usually encountered in small groups feeding on the ground. When flushed the wings produce a dry rattle and the birds usually take refuge on a cliff face or rock pile. When displaying (which involves bobbing of head downwards with neck stretched out) and when perched, occasionally makes a soft purring croak reminiscent of the more familiar call of the Croaking Ground-Dove of the Peruvian coast. Nests on the ground or cliff faces, also holes in buildings. Can be seen at the start of the Inca Trail near Llactapata ruins, throughout the Sacred Valley and the Cusco area including rooftops in the city itself. Named for Cécile Gautrau, daughter of French naturalist René Lesson.

Black-winged Ground-Dove   *Metriopelia melanoptera*
21-23 cm. 2000-4300 m. In flight shows dark wings with sooty gray linings and white wrist spot. Vent and longish square tail black. Found in arid and semi-arid regions. May feed far from trees, but gathers in *Polylepis* and *Eucalyptus* in the late afternoon. Also, on hillsides with scrub and cactus, agricultural terraces and fields. Usually in small flocks feeding on the ground. Will perch in trees and bushes. The seldom heard call is a rolling *'treee-ooi'* delivered from a bush or rock. Can be seen in the Cusichaca Valley within the Sanctuary, the Sacred Valley and Cusco areas.

CUCKOOS AND ANIS   Cuculidae

A cosmopolitan family whose members are similar, but differ dramatically in behavior. They are slim-bodied, long tailed birds, many with secretive habits. Mostly found in the tropics, they move actively, like passerines, hopping from branch to branch.

Squirrel Cuckoo   *Piaya cayana*
43 cm. < 2500 m. A large, lanky cuckoo. Found in dry to wet forest, forest borders, tall secondary growth and semi-open areas with trees. Encountered alone or in pairs in thick vegetation and vine tangles where they run and glide with squirrel-like movements or strong hops. Often seen gliding across clearings or from tree to tree. The most frequently heard call is an explosive *'chic'* or *'skiik-whaah.'* Mostly a lowland species but ascends to 2500 m. Rare in our area. Can be seen along the Urubamba River near Puente Ruinas railway station.

Dark-billed Cuckoo   *Coccyzus melacoryphus*
25 cm. < 2400 Vag to 3400 m. Found in a variety of habitats, including forest edge, gardens, dry bushy slopes, secondary growth and quite dry xerophytic areas. Found quite low in thick leafy vegetation, usually alone, but sometimes in pairs. Sits quietly looking for caterpillars and other insects. Hops and flies through thickly interlacing vegetation with surprising ease. The seldom heard song is a descending series of *'coos.'*

Yellow-billed Cuckoo   *Coccyzus americanus*
26-29 cm Vagrant 2000 m. Boreal migrant mostly to the eastern lowlands in Peru, but a vagrant recorded at Machu Picchu at 2000 m. Note white underparts, rufous webs of remiges and yellow base to lower mandible. Rare vagrant.

Greater Ani   *Crotophaga major*
46-48 cm. Vagrant 2000 m. One was present in the grounds of a hotel at Aguas Calientes 11-17 April 2009. Normally very social and in groups of up to 20, almost always associated with water. Makes electronic like grunts, hisses, groans and bubbling sounds. Movements little understood but is known for vagrancy and, where resident, often disappears for periods of time – migrant? Very rare wanderer in our region.

Smooth-billed Ani *Crotophaga ani*
33 cm. < 2800 m. Tail long and loose. Found in bushy pastures, clearings and open areas in humid forest, corn crops and riverine thickets. Colonizes deforested areas rapidly, especially as montane forest is cleared. Very gregarious and usually found in small, loose groups. They perch conspicuously and are easy to observe, perching on small trees, bushes, fence posts or the ground. The flight is weak and they seldom fly far – usually a few flaps and then a sailing glide. The often given call is a rising *'oouuuu-eenk'* often given in flight. Mostly a lowland species, but has colonized upwards as a result of deforestation. Occurs up to 2800 m. Uncommon at Machu Picchu and best looked for in the lower areas near the Mandor and Aobamba Valleys. A few records from near Cusco at 3350 m.

BARN OWLS   Tytonidae

Only one species occurs in the Americas. They differ from typical owls by the heart-shaped facial disk and long legs. They hunt rodents and are faithful to nest sites.

Barn Owl *Tyto alba*
38 cm. 1800-4000 m. In flight looks unmarked and almost totally white. Found around human habitation, agricultural areas and lightly wooded and open terrain. Nests in rock crevices, caves and hollow trees. Most active at dusk and at night, gliding buoyantly and silently over open country in search of small rodents. Perches on dead stumps and fence posts. Rocks head and body violently from side to side when alarmed. Has a variety of hissing and screeching calls but does not hoot. Can be found at all elevations but is inexplicably uncommon in our area. Can be seen at Huacarpay Lakes south of Cusco.

TYPICAL OWLS   Strigidae

Owls are easily recognized by their compact shape, wide heads with broad facial disks and soft cryptic plumage enabling silent flight. They have incredible eyesight and binocular vision. The eyes are fixed and to compensate for this they can swivel their heads 270 degrees. They have well developed hearing with facial disks that help concentrate sound. Owls hunt mostly at night, but they can see well during the day and some species will hunt in the daytime. Their prey, located by sight and sound, is usually small mammals, birds, insects and sometimes frogs. Prey is swallowed whole, the fur and bones are regurgitated in the form of pellets. Most owls nest in holes in trees, rock crevices and old stick nests of other birds. They lay white eggs. Due to their nocturnal habits they are seldom seen and little known.

White-throated Screech-Owl *Megascops albogularis*
25 cm. 2000-3000 m. A very dark owl with almost no ear tufts and a contrasting white throat. Inhabits humid montane and pre-montane forest, mostly open forest with a broken canopy, forest edge and glades, often where there is bamboo. Strictly nocturnal mostly feeding on large insects in the canopy. Nests in tree holes or old nests of other birds. The song is a series of 5-6 fluid hoots – *'bu-bu-bu-bu-bubu.'* Can be seen and heard near Wiñay Wayna ruins along the Inca Trail.

Band-bellied Owl *Pulsatrix melanota*
37 cm < 2300 m. Uncommon in humid pre-montane forest. Note distinctive facial markings and rufous bands on the underparts. Song a series of muffled hoots, often compared with wobbling a thin sheet of metal. Also mewing calls. Best looked for between Aguas Calientes and the Mandor Valley.

Great (Lesser) Horned Owl *Bubo (magellanicus) virginianus*
48-56 cm. > 3000 m. Some authorities consider this taxon to be conspecific with the Great Horned Owl *(Bubo virginianus)*. König *et al.* (1999) and Jaramillo (2003) proposed that *magellenicus* be elevated to species rank based on differences in vocalizations, morphology and DNA, however, further studies of intermediate populations are needed . A very big owl with yellow eyes and conspicuous ear-tufts. Found in open or wooded terrain, sometimes near human habitation or on rocky hillsides. Crepuscular when hunting, but will also hunt during the day, mainly for small mammals. Drops on prey from an elevated perch levelling off just above the ground. Nests in caves. Gives two deep, low pitched hoots followed by a purring sound *'hu-Oohh-urrr.'* Also screams loudly. Can be seen in the evening near Llactapata ruins at the start of the Inca Trail and throughout the Sacred Valley of the Incas and even in Cusco city.

Smooth-billed Ani

Barn Owl

White-throated Screech-Owl

Band-bellied Owl

Great (Lesser) Horned Owl

Rufous-banded Owl   *Ciccaba albitarsus*
35 cm. 1900-3500 m. A compact owl without ear tufts, rounded head. Found in humid montane and pre-montane forest. Strictly nocturnal and most active (vocalizing) just after dark and pre-dawn. Feeds on small mammals and large insects, mostly in the canopy. The song is a fast series of hoots *'hu hu-hu-hoo-aaa'* with a pause after the first note. Can be seen and heard along the Inca Trail at the Wiñay Wayna ruins.

Yungas Pgymy-Owl   *Glaucidium bolivianum*
16 cm. 1800-3000 m. A tiny owl with a rounded head. The Yungas Pygmy-Owl is very similar to the Andean Pygmy-Owl *(Glaucidium jardinii)*, which it replaces in southern Peru. Has two main color morphs: brown (common) and rufous (rare). Found in humid montane and pre-montane forest with a predominance of alders *(Alnus jorullensis)*, thick moss and epiphytes. Nocturnal and diurnal particularly in cloudy weather. Feeds on insects and small birds, usually hunting at mid to high levels in the canopy. Nests in holes in trees, often old woodpecker holes. The song is a slow series of well spaced, whistled notes. Fairly common within the Machu Picchu Sanctuary and at Abra Malaga.

Peruvian Pygmy-Owl   *Glaucidium peruanum*
16 cm. < 3400 m. Probably more than one species involved: the intermontane form in our region may be distinct from the coastal form. Gray morph: dark gray-brown to slaty brown above. Rufous morph: warm rufous brown above. Found in semi-arid woodland, *Eucalyptus* groves, riparian thickets and hedgerows near agricultural land as well as gardens. Nests in holes in trees and feeds mostly on large insects. The song is a long series of well spaced notes with a falling pitch *'boop boop boop,'* etc. Fairly common in the Cusco Valley and Sacred Valley of the Incas.

Burrowing Owl   *Athene cunicularia*
24 cm. 3000-4000 m. The *juninensis* race is present in our area. Found in a variety of open habitats, mostly treeless plains and agricultural fields. Mostly diurnal. Nests in burrows. Mainly terrestrial, but perches on rocks, walls and posts. Feeds on small mammals, large insects, beetles and lizards. Bobs up and down when nervous. Uncommon but can be seen near the town of Maras on the Chinchero Plateau and occasionally near Huacarpay Lakes. In Greek mythology, Athena, for whom the owl was sacred, was the goddess of wisdom and war.

OILBIRD   Steatornithidae

A single species related to nightjars. Oilbirds are the only nocturnal fruit eating birds in the world. They roost and breed in colonies in caves from northern South America where it is most numerous and locally along the Andes as far south as Bolivia.

Oilbird   *Steatornis caripensis*
48 cm. 1800-2000 m. Like a nightjar, but much larger with a strong hawk-like bill. Eye-shine red. Roosts during the day in caves in humid pre-montane forest. Forages over forest at night. Feeds entirely on fruits, mostly palm fruits, by hovering and plucking from the trees at night. Swallows fruits whole and regurgitates seeds. Flies up to 150 km when foraging. Mainly at altitudes below 3000 m. Has been recorded several times at Machu Picchu near the town of Aguas Calientes and once near Cusco City (vagrant or wandering juveniles?). Breeding caves could exist near Machu Picchu but have yet to be discovered, if indeed there are any. The name comes from the Greek: *steatos* = fat, *ornis* = bird. Baron von Humboldt recorded that fat young oilbirds were collected each year and their fat melted down to extract oil.

Rufous-banded Owl

Yungas Pgymy-Owl

Amazonian-slope bird

Peruvian Pygmy-Owl

Pacific-slope bird

Burrowing Owl

Oilbird

NIGHTHAWKS AND NIGHTJARS Caprimulgidae

A cosmopolitan family better identified by their songs than by sight. Almost totally nocturnal, these birds are rarely seen in daylight except when flushed from cover. Nighthawks have long pointed wings and fly high at dusk and dawn. Nightjars mostly occur closer to the ground and have shorter rounded wings and long tails. They have cryptic plumage and camouflage themselves by sitting motionless among leaf litter. They lay two eggs on the ground and the young are fed for a long time before fledging.

Rufous-bellied Nighthawk *Lurocalis rufiventris*
25 cm. 1800-3300 m. At rest wings project well beyond tip of tail. Found in humid montane and pre-montane forest, forest edge and secondary growth. Hawks erratically with raised wings for insects above the canopy. Alone or in small groups at dawn and dusk. The call is an evenly spaced series of *'kwaa'* notes or mellow hoots falling in pitch.

Nacunda Nighthawk *Chordeilis nacunda*
27-30 cm. Vag. 3000 m. Austral migrant to the Amazonia lowlands with one record from Huacarpay Lakes south of Cusco (photographed) May 13th 2006 (Venero 2008). Easily identified by large size, contrast between dusky breast and white belly and large white wingband on broad round tipped wings. Vagrant.

Band-winged Nightjar *Systellura longirostris*
24 cm. 2400-4400 m. The *atripunctatus* race is present. Found in rather dry situations, open woodland and scrubby slopes, puna grassland and groves of *Polylepis*. Nocturnal. Feeds by short sallies from a low perch for flying insects. Sings shortly after dusk – a high pitched *'cheet-wit'* repeated at intervals, also a tiny buzzy *'zuueeert'* rising toward the end. Can be seen around Huacarpay Lakes.

Swallow-tailed Nightjar *Uropsalis segmentata*
22 cm (not including male's streamers). 2500-3600 m. A very dark nightjar with black wings and long forked tail. Adult male shows scissor-like streamers, twice the body length. Found in humid montane and elfin forest where there are clearings, forest edge, bushy slopes and at tree line. At dusk and pre-dawn, hunts for flying insects over grassy slopes and along forest edges, also sallies from low perches for mostly moths. Song is a long drawn out *'purrrrr sweeeeeeeerrrrrr'* delivered from a low perch or on the ground. Can be seen at dusk near Sayacmarca ruins along the Inca Trail and at tree line at Abra Malaga.

Lyre-tailed Nightjar *Uropsalis lyra*
25-28 cm (not including male's streamers). 1800-2100 m. Similar to the preceding species, but a slightly larger bird with longer streamers – almost three times the body length in the male, curved into a lyre-shape. Inhabits humid montane and pre-montane forest, and generally replaces the preceding species at lower elevations. In clearings and glades, almost always near cliff faces or rocky ravines. Feeds on the wing or sallies from a low perch for flying insects. Nests on ledges on cliff faces, road cuts etc., often close to the ground. Has a communal display. Song is a wood-quail-like *'tree-couee tree-couee tre-coueee'* repeated 4-5 times. Can be seen at dusk near the hot springs at Aguas Calientes and along the railroad track, often near the Machu Picchu Pueblo Hotel.

SWIFTS Apodidae

A cosmopolitan family that vaguely resembles swallows, but swifts have sickle shaped wings adapted for high speed sailing. They are mostly shades of brown and black and have small bills, but very large gapes. They are the most aerial of all birds, feeding on insects, mating and even sleeping in the air. Some species roost on vertical cliffs or in tree-trunks. The paired toes oppose each other to provide a powerful grip and, in some species, stiffened tails give additional support when the bird is roosting. Swifts are very social and found in small to very large groups. They can enter torpor to save energy. Nesting is usually semi-colonial with tiny nests glued with saliva to vertical rock faces. They lay 4-6 eggs and have a long nesting cycle.

Chestnut-collared Swift *Streptoprocne rutilus*
15 cm. < 3000 m. Longer tailed and narrower winged than *Chaetura* swifts, tail square or slightly notched. Has a steadier, less erratic flight. Found in mountainous humid pre-montane forest, flying over more open terrain and grassy ridges. Usually encountered in single species flocks but will join flocks of *Chaetura* swifts, flying high or down valleys. Roosts in tree trunks. Call is high pitched *'chittering'* and *'buzzing.'* Found at elevations of up to 3000 m but mostly below 2300. Uncommon at Machu Picchu.

White-collared Swift *Streptoprocne zonaris*
20 cm. 1800-4000 m. Unmistakable, large black swift with a broad white ring around the neck. Shows an obvious forked tail. Found in almost all habitats except puna, in mountainous country from humid forested to dry inter-montane valleys. A highly social species found in flocks of hundreds or more. They feed on insects at all heights depending on weather conditions, covering large areas during daily foraging. Soars with wings bowed, but flight profile depends on the birds altitude. They roost and nest in caves mostly behind waterfalls. Very vocal: calls include nasal twitters *'chee-chee-chee,'* *'whiss-whiss,'* or *'cheet-cheet'* – often many birds together sounding like a distant parakeet flock. Can be found at all elevations and can be seen virtually anywhere in our area.

Chimney Swift *Chaetura pelagica*
13 cm. 2300-4500 m. A boreal migrant present in Peru from November to April. In flight shows a thick "cigar-shaped" body with a short squared off tail and projecting spines at close range. Note short tailed appearance and fluttering flight, with rapid wing-strokes and much soaring. Mostly found in open terrain and very often in towns and villages. Highly gregarious. Mostly winters on the Peruvian coast but has been recorded as a vagrant at Machu Picchu and the Cusco Valley. Rare.

White-tipped Swift *Aeronautes montivagus*
12 cm. 1800-2400 m. In flight similar to preceding with sickle-shaped wings and "cigar-shaped" body, but with slightly forked or notched tail (not always apparent). Found in humid to semi-humid areas, usually over forested hills and gorges, bushy slopes and ridges. Gregarious, usually found in small groups flying at medium heights. Quite vocal – a long buzzing, clicking trill or high pitched squeaking. Quite common at Machu Picchu ruins and around the Wayna Picchu peak.

Andean Swift *Aeronautes andecolus*
12 cm. 2500-4500 m. The *peruvianus* race is present in our area. In flight is longer winged and tailed than the preceding species, the tail looking narrow and slightly forked. Found in semi-arid mountainous country over bushy slopes and also around towns. Gregarious and often fly high along precipitous cliff faces in flocks or small groups. The call is a shrill *'zeezeezeeezeer'* often given by more than one individual. Can be seen at the start of the Inca Trail near Llactapata ruins and in the Sacred Valley of the Incas.

HUMMINGBIRDS   Trochilidae

A strictly New World family, hummingbirds include some of the smallest birds in the world. They are adapted to extract nectar from flowers while in flight and have specially designed bills and tongues to do so. While most species hover to extract nectar, many will also perch to feed. Flexible shoulder joints permit backwards flight. Wingbeats (22-78 per second) are extremely rapid, producing a humming sound. Hummingbirds have brightly colored, iridescent and glittering plumages. Some are aggressive and hold large territories; others defend a blooming tree or trap-line – visiting particular flowers on a regular route. Hummingbirds are important pollinators of many flowers. Most hummingbirds take insects as well as nectar. Some species found at higher elevations become torpid at night to save energy. Males sing from exposed perches and at leks. The female alone builds the nest, incubates and raises the young. Nests are built on thin twigs or in crevices; sometimes a hanging woven structure is made. Two whitish eggs are laid.

Buff-tailed Sicklebill   *Eutoxeres condamini*
16 cm (incl. 2.7 cm bill) < 2500 m. Exceptional curved sickle shaped bill. The *gracilis* race is present at Machu Picchu. Found in humid pre-montane forest but sometimes higher into montane forest, overgrown clearings, forest edge, along streams and in thickets of *Heliconia* sp. Retiring and difficult to see. Feeds by trap-lining for nectar and will take small insects. The nest is attached to the underside of a long hanging leaf with spiderwebs. Uncommon. Named for Charles Marie de la Condamine (1701-1774) French scientists and mathematician.

Green Hermit   *Phaethornis guy*
13 cm (incl. 4 cm bill). < 2300 m. Bill long and decurved. The *apicalis* race is present at Machu Picchu. Found in the understory of humid pre-montane forest, adjacent forest edge, well developed secondary growth and overgrown gardens. Feeds by trap-lining for nectar between varieties of flowers – will also take spiders from webs. The nest is a hanging cone-shaped affair attached to the underside of a long leaf in the understory. Call is a loud *'tsweep.'* The song at the lek is a nasal *'heweet heweet hewwet'* repeated for long periods. Can be seen in the lower sections of the MPHS.

Green-fronted Lancebill   *Doryfera ludovicae*
13 cm (incl. 3.5 cm bill). 1800-2500 m. Usually sits with long thin bill upturned. A trap-lining hummer visiting flowers with long corollas, often hovers for insects. Found in the understory and middle story of wet ravines, often close to forested waterfalls and water seeps in hillsides in rocky situations. Calls include a thin *'tsip'* and *'tsee'* often in series.

Green Violetear   *Colibri thalassinus*
11 cm (incl. 1.8 cm bill). < 2400 m. Bill slightly curved. The *crissalis* race is present in our region. Found on open, humid, shrubby mountain slopes and pastures with scattered trees, overgrown landslides and forest edge. Feeds on a variety of flowers at mid-levels. Male often sings from an exposed perch at mid-heights. The song is a high pitched *'tsup-chip chip-tsup tsup-chip,'* etc, repeated endlessly at all times of day. Aggressive, display flight is low and undulating. Can be seen on the bushy, open slopes along the Inca Trail between Intipunku and the Machu Picchu ruins. Quite common, attends feeders at the Machu Picchu Pueblo Hotel. Greek: *thalassinus* = Sea Green.

Sparkling Violetear   *Colibri coruscans*
14 cm. (incl. 2.5 cm bill). 2000-4000 m. Bill slightly curved. A rare melanistic form occurs, though we have not seen this in our region. Perhaps the most common and widespread hummingbird of the Peruvian Andes. Inhabiting a variety of mostly not-too-humid habitats, it is catholic in its tastes. Non-breeders will wander into humid forest areas and congregate at flowering trees, particularly *Erythrina* sp. Prefers all kinds of dry open areas with scattered trees and shrubbery, secondary growth and gardens. Readily adapts to *Eucalyptus*. Feeds at all levels on nectar from a wide variety of flowers, hawks for insects. Territorial, aggressive and dominant over other hummingbirds. Commonly seen display flight consists of the male ascending about 10 m from its exposed song perch and descending same way twittering with a fanned tail. The song, delivered from an exposed perch, is an endlessly repeated *'tzrrt-tzrrt-tzrrt.'* Common throughout the Cusco area, the sacred Valley and near Wayllabamba along the Inca Trail.

Amethyst-throated Sunangel *Heliangelus amethysticollis*
11 cm. (incl. 1.8 cm bill) 1800-3200 m. Rather short straight blackish bill. Inhabits humid montane and pre-montane forest with lots of moss and epiphytes, also bushy slopes with a few tall trees and damp bushy ravines. Mostly at mid-levels, but will feed lower at forest edge. Often perches while feeding and defends concentrations of nectar producing flowers. Hawks for insects from a favorite perch, may associate with mixed feeding flocks. On alighting, characteristically holds its wings aloft for a few seconds like an angel. Calls include a dry *'trrr.'* A conspicuous hummingbird which can be seen near the Wiñay Wayna ruins along the Inca Trail and at Abra Malaga.

Speckled Hummingbird *Adelomyia melanogenys*
10 cm (incl. 1.4 cm bill) 1200-2800 m. Bill straight and short. Throat finely dotted with dusky and sometimes a little green. Found in humid pre-montane forest, forest edge and along watercourses. Forages alone in the understory, sometimes to mid-levels. Often sits with slightly lowered wings. Feeds on the nectar of various flowers and small insects. Sometimes pierces the base of flowers or uses existing holes made by flowerpiercers to obtain nectar. Calls are typically high pitched *'zit'* or *'zwee.'* Fairly common along the Urubamba River at Machu Picchu.

Long-tailed Sylph *Aglaiocercus kingi*
10-19 cm (depending on tail length incl. 1.4 cm bill). 1800-2700 m. Short black bill. Tail narrow and deeply forked with 12 cm lateral streamers, iridescent violet blue-green above and bluish-black below. Found in humid montane and pre-montane forest with many vines and a rather open canopy. Feeds mostly in the crowns of trees on nectar and insects by defending territory or trap-lining. Also pierces base of corollas for nectar. Hovers at flowers but sometimes clings to them. Will hawk for insects from a perch. Bobs tail when perched. Calls include frail *'tzit'* and *'trrt'* notes. Can be seen along the railway track between Aguas Calientes and the Mandor Valley and attends feeders at the Machu Picchu Pueblo Hotel.

Andean Hillstar *Oreotrochilus estella*
14 cm (incl. 2 cm bill). 3400-4600 m. The nominate race is present in our region. Mostly found on high puna grasslands, *Polylepis* woodlands. Favors areas where there are bushy ravines and rocky outcrops. Also, bushy gardens and around houses. Aggressive. Often perches on the top of bushes for long periods. Flies fast and high over terrain. Feeds on low bushes and shrubs by clinging as it takes nectar. Also, hawks for insects and gleans vegetation. In the breeding season (September to December), males wander widely across open terrain and females hold territory in bushy ravines. They roost in well protected crevices and caves and have the ability to go into torpor. Calls include fine *'tij'* and *'zeer'* notes. Can be seen near Pacasmayo along the Inca Trail, higher side valleys in the Urubamba Valley and at Abra Malaga.

Black-tailed Trainbearer *Lesbia victoriae*
15-26 cm (incl. 11-18 cm tail). 2700-4000 m. The *berlepschi* race is present at Machu Picchu. Bill short and black, slightly de-curved, especially in the male. Note more V shaped gorget than the following species. Found in less humid bushy ravines, bushy slopes, and hedgerows, gardens and *Polylepis* woods. Aggressive and territorial. Forages on a variety of flowers, including introduced *Eucalyptus* at mid to high levels. When perched often bobs tail. Hawks and gleans insects. Has a high aerial display flight with spread streamers. Calls include *'tic'* notes and a thin *'zeet.'* Song is *'ti-ti-tit- trrrr tic-tic-tic.'* Can be seen throughout the Sacred Valley of the Incas and Cusco area. From the Greek: Lesbia – a woman of Lesbos, Victoriae – named for Victoire Mulsant mother of French ornithologist Martial Mulsant.

Green-tailed Trainbearer *Lesbia nuna*
15.5-18 cm (male incl. tail of 10.5-12 cm); female 11.6 cm (including tail of 5.5 cm). 2000-3800 m. Similar to above, but long forked tail blackish with green tips. Note rounded gorget compared to previous species. Inhabits bushy slopes and ravines, semi-humid scrub and gardens. Usually in somewhat more humid habitat than the Black-tailed Trainbearer. Aggressive, but submissive to the preceding species. Feeds on a variety of flowering shrubs and occasionally hawks for insects. Display flight different from Black-tailed Trainbearer – male flies in a zigzag pattern in front of perched partner. Calls include *'bzeeet'* and a high pitched *'seep.'* Song is a canastero-like *'zee-zee-zee-zzee-zeeet.'* Can be seen throughout the Sacred Valley of the Incas and Cusco area, also near Wayllabamba along the Inca Trail.

Purple-backed Thornbill *Ramphomicron microrhynchum*
8.5 cm. 2500-3650 m. Bill short, straight and black (the shortest bill of any Hummingbird). Inhabits the borders of humid montane forest, elfin forest and nearby bushy slopes. Flight is bee-like. Feeds by trap-lining for nectar at mid to high levels on the outside of vegetation. Often uses holes made by flowerpiercers in corollas. Also, hawks for insects. The song is along weak series of notes *'ti-ti-ti-ti.'*

Rufous-capped Thornbill *Chalcostigma ruficeps*
10 cm (incl. 1.1 cm bill). 1500-3600 m. Bill short, straight, slightly upturned at the tip, and black. Found in humid montane and pre-montane forest, forest edge and borders. Also, in well-developed humid secondary growth. Forages at low to medium levels on a variety of flowers and will make holes or use existing flowerpiercer holes at the base of corollas to extract nectar. Clings to flowers rather than hovering. Takes insects to supplement diet of nectar. Song is a soft frail trill. Flight call *'tzee-zee-zee.'* Found at elevations of between 1500 m and 3600 m but more frequently around 2500 m. Generally uncommon.

Olivaceous Thornbill *Chalcostigma olivaceum*
12-14 cm. (incl. 1.2 cm bill). 4000-4600 m. Bill short, straight and black. Female has partly disintegrated beard and pale tips to the outer tail feathers. Inhabits puna grassland and moist valley bottoms in high Andean valleys. Also, edge of *Polylepis* groves and bushy slopes at high altitude. Feeds on nectar and insects, specializes on prostrate small flowers on the ground. Unique behavior, feeding on flowers and insects while walking across the ground, sometimes hovering close to the ground and darting up after flying insects. Seasonal crowding occurs at concentrated nectar sources. Can be found around the Salcantay massif and at Abra Malaga.

Blue-mantled Thornbill *Chalcostigma stanleyi*
12-13 cm (Bill 1.2 cm.) 3600-4400 m. Bill as above. Found on steep rocky slopes with *Gynoxys* and *Polylepis* woods or other scrub, occasionally to surrounding open rocky terrain. Feeds on a variety of flowers close to the ground, also feeds on insects and sugary secretions on the underside of *Gynoxys* leaves. Feeds whilst clinging to vegetation with much fluttering. Occasionally hawks for insects. Call a weak *'djeer.'* Can be seen near Llulluchapampa along the Inca Trail and in *Polylepis* at Abra Malaga. Named for Edward Stanley 15th Earl of Derby (1826-1893).

Bearded Mountaineer *Oreonympha nobilis*
Peruvian Endemic. 14-17 cm, depending on tail length, (incl. 2.2 cm bill). 2500-3900 m. The nominate race is present. Inhabits dry Andean Valleys with scrubby slopes and open woodland, alder lined ravines, and agricultural land with *Nicotiana* bushes and *Eucalyptus* trees. Also, dry slopes with columnar cactus. Usually found near rocky outcrops. Avoids moist situations. Rather shy and submissive to other hummingbirds. Feeds on nectar by both hovering in horizontal position below the flower incessantly opening and closing tail, or by clinging. Also, hawks for insects. Can be seen at the start of the Inca Trail near Llactapata ruins, throughout the Sacred Valley and around Huarcarpay Lakes south of Cusco.

Tyrian Metaltail *Metallura tyrianthina*
9 cm (incl. 1.3 cm bill) . 2500-3600 m. The *smaragdinicollis* race is present in our region. Tail in both sexes is deep blue. Found in lighter humid montane forest, elfin forest, secondary growth and bushy clearings at tree line. Also, bushy ravines in semi-arid areas, often with alders (*Alnus* sp.) Mostly encountered at low to mid-heights hovering or clinging to flowers. Occasionally pierces base of corollas to extract nectar. Hawks for insects. Makes variety of harsh *'shric'* and *'trrr'* notes, and *'zee-tee titititit-zwee dwee-zeee,'* etc. Also *'zii-zii-zii-zii'* delivered from a low perch. Quite common near tree line throughout the MPHS, Abra Malaga and moister side valleys of the Urubamba River. Named after the color Tyrian purple. This dye was first produced by the ancient Phoenicians in the Lebanese port of Tyre (hence the name). Variously known as Royal purple, Tyrian purple, or purple of the ancients; it was produced from the mucus of a sea snail in the genus *Murex*. Highly valued by the Ancients and prized by the Romans, its production declined with the decline of the Roman Empire.

Purple-backed Thornbill

Rufous-capped Thornbill

Olivaceous Thornbill

Blue-mantled Thornbill

Bearded Mountaineer

Tyrian Metaltail

Scaled Metaltail   *Metallura aeneocauda*
11 cm (incl. 1.5 cm bill). 2800-3600 m. Inhabits bushy clearings in humid montane and elfin forest, forest edge and nearby bushy slopes. Feeds within a few meters of the ground on a variety of flowers with 2-4 cm corollas often clinging to the flowers. Also, takes insects. Call/song is a buzzy series of notes *'zew-zew-zew-zwhizizi-zwihzizi.'* Can be seen near Sayacmarca ruins along the Inca Trail and but best seen at tree line at Abra Malaga.

Buff-thighed Puffleg   *Haplophaedia assimilis*
9-10 cm. 1800-2500 m. Conspicuous buffy white leg puffs. Inhabits the undergrowth of humid montane and pre-montane forest, forest borders and edge. Territorial and aggressive. Holds territories around clumps of flowers with short corollas. Stays mostly low in the vegetation only reaching mid-levels when *Inga* sp. trees are in flower. Gleans vegetation for insects on occasion. Best looked for in the Mandor and Aobamba drainages.

Sapphire-vented Puffleg   *Eriocnemis luciani*
12-14 cm incl. 2.2 cm bill. 2400-3500 m. Female similar but shows a white stripe on the belly. Inhabits humid montane and pre-montane forest, mostly at the edge, also elfin forest. Will venture onto bushy slopes. Territorial and defends concentrations of flowers. Mostly feeds at low to mid-levels, often clutching flowers while extracting nectar rather than hovering. Will land on the ground for terrestrial flowers and hawk for insects. Can be seen along the Inca Trail between Wiñay Wayna and Intipunku and tree line at Machu Picchu and Abra Malaga. Named for French etntomologist Lucien Buquet (1807-1889).

Shining Sunbeam   *Aglaeactis cupripennis*
12-13 cm. (incl. 1.5 cm bill). 2500-4000 m. Female is less iridescent on the back. Inhabits fairly humid to semi-arid bushy slopes admixed with alders (*Alnus* sp.), secondary growth, also sometimes *Polylepis* woodland. Perches conspicuously on exposed treetops and glides downhill on spread wings. Often holds wings out in a V or flaps them slowly after alighting. Feeds perched, often with raised wings, on nectar from flowers and insects. Also, hawks for flying insects. Calls include high pitched descending *'tzee-tzee-zee-zee,'* etc. or *'tzerr-tzerr.'* Can be seen in the Pacasmayo Valley along the Inca Trail and the south side of the Abra Malaga pass.

White-tufted Sunbeam   *Aglaeactis castelnaudii*
Peruvian Endemic. 12 cm (incl. 1.8 cm bill). 2600-4100 m. Tuft of white feathers on the central breast. Habitat similar to the preceding species – rather open shrubby hillsides with some taller trees mixed with alders (*Alnus* sp.). Also, in glades in semi-arid forest and *Polylepis* groves. Generally at higher elevations than the Shining Sunbeam, but where they occur together, this species is submissive. Feeds on nectar by clinging to flowers and hovers for flying insects. Call include *'py-py-py'* and very high pitched *'zeets.'* Can be seen at Llulluchapampa along the Inca Trail and the south side of Abra Malaga pass. Named for the splendidly named Frenchman, Francois Nompar de Caumont Laporte Compte de Castelnau (1810-1880): diplomat, explorer and collector in tropical America.

Bronzy Inca *Coeligena coeligena*
14 cm (incl. 3.6 cm bill). 1800-2200 m. Lower mandible sometimes shows some yellow at the base. Found at the edge of humid pre-montane forest, and in forest clearings, edges of trails etc. Mostly forages by trap-lining for nectar at the edge of, but also inside forest. Mostly at mid to lower levels, but will go into the crowns of flowering trees. Will take small insects by hover-gleaning or hawking. Can be seen in the Mandor Valley.

Collared Inca *Coeligena torquata*
14.5 cm (incl. 3 cm bill). 1800-3000 m. The *omissa* ("Gould's Inca") race is present in our region and this with *C. inca* has been considered a separate species. Inhabits humid montane and pre-montane forest and forest borders. Feeds at low to mid-heights sometimes in low canopies mostly hovering along the edge of dense vegetation, trap-lining for nectar. Hawks for insects. Flies fast through the forest and reveals itself by flashing the white of the tail periodically. Note cinnamon breast band and white tail. Can be seen along the railway track between Puente Ruinas railway station and the Mandor Valley. Attends feeders at the Machu Picchu Pueblo Hotel.

Violet-throated Starfrontlet *Coeligena violifer*
14 cm (incl. 3 cm bill). 2500-3900 m. Found at the edge of, and in clearings in, humid montane forest as well as elfin forest. Feeds by trap-lining in low strata or along the edge of bushes, occasionally along forest trails, will go into the canopy. Nuptial display is a butterfly like pendular flight above a perched partner. Call is a series of *'zwit'* notes. Can be seen along the Inca Trail between Sayacmarca and Phuyupatamarca and at Abra Malaga.

Mountain Velvetbreast *Lafresnaya lafresnayi*
11.5 cm (incl. 3 cm bill). 1800-2300. Bill thin, black and de-curved. Found along the borders of humid montane and pre-montane forest, bushy slopes and overgrown landslides. Forages at low levels with fluttering flight whilst spreading and closing tail. Males hold territory on nectar-rich flower patches whilst females trap-line for nectar. Also hawks for small flying insects. Calls include a whistled *'zeee'* and soft *'tzrr'* notes.

Sword-billed Hummingbird *Ensifera ensifera*
22 cm. (incl. 8-10 cm bill). 2400-3600 m. Bill extremely long, slightly upturned. Much individual variation in bill length, but the longest bill of any hummingbird. Found in humid-montane, semi-humid forest, elfin forest and forest edge with many vine tangles. Holds bill upright. Trap-lines for nectar at mid to high levels on the outside of trees and shrubbery. Specializes on pendent flowers with long corollas, mostly hovering whilst extracting nectar but sometimes feeds whilst perched. Will also hawk for insects in a swift-like manner. Voice is plaintive whistle, also a guttural *'trrr.'* Can be found along the Inca Trail between Wayllabamba and Warmiwañusca pass and at Abra Malaga. Fairly common.

Great Sapphirewing *Pterophanes cyanopterus*
19 cm (incl. bill 3 cm) 2600-3600 m. Large. Found in open montane forest, forest edge, elfin forest and open puna with a few scattered plants. Sometimes in drier, semi-humid habitat with alders. Flies very fast with erratic wing-beats and glides. Feeds on nectar from flowers on the outside of thickets and bushy tangles, with heavy, hovering wing beats, sometimes clutching to flowers. Insects are caught by hawking. Sometimes associates with mixed species flocks at tree line. Call is a piercing high pitched *'zreeee,'* also an agitated *'tittitzerrr.'* Can be seen along the Inca Trail near Runkuracay and at Abra Malaga.

Chestnut-breasted Coronet *Boissonneaua matthewsii*
11.5 -12 cm (incl. bill 1.8 cm) 1800-3300 m. Found in humid pre-montane forest. Inhabits the interior and canopy, but sometimes seen at forest edge. Defends patches of flowers (and hummingbird feeders) often clinging to flowers. Will also hawk for insects. Common on the grounds of the El Pueblo Hotel in Aguas Calientes and attends feeders aggressively. Named for English botanist and collector Andrew Matthew (d.1841).

Booted Racket-tail *Ocreatus underwoodii*
Male 11-15 cm (incl. elongated tail feathers), female 7.5-9 cm. 1800-2400 m. The *annae* race is present at Machu Picchu. Conspicuous tawny leg puffs. The male's distinctive long tail has bare shafts which are crossed and end in 1 cm wide steel blue rackets. Found in humid montane and pre-montane forest, also well-developed secondary growth. Can be seen in the understory and at forest edge from low to high levels. Flight is wavering and bee-like. Holds wings outstretched for a few seconds after alighting. Mostly feeds in the canopy on a wide variety of tubular flowers, holds territories. Clings to flowers while feeding. Makes some frail twittering notes, the wings make a distinctive hum. Can be seen in the *Inga* trees at the bottom of the Mandor Valley and attends feeders at the Machu Picchu Pueblo Hotel. Specimens of this species existed in various London cabinets, but it was unknown to science. A drawing was sent to Lesson in 1832 by Mr. Underwood on behalf of Charles Stokes, a London stockbroker and collector.

Fawn-breasted Brilliant *Heliodoxa rubinoides*
12 cm. < 2300 m. Uncommon. Forages at mid-levels and understory in humid pre-montane forest. Buffy below with variable spotting and small pink gorget. Song a long series of *'tchew'* notes. Only recorded so far form the Urubamba Valley near Aguas Calientes.

Giant Hummingbird *Patagona gigas*
23 cm (incl. 3.5 cm bill). 2500-4000 m. Bill straight and rather thick. The world's largest hummingbird. Wings long and narrow; resembles a swift or bee-eater in flight. Females show some dusky spotting below. Found in open arid habitats with low bushes and some trees and on slopes with bushes, columnar cactus, agaves and thistles, also hedgerows and gardens near houses. Visits small copses. Avoids moist forest. Generally territorial and aggressive, chasing violetears and hillstars. Flies with erratic wingbeats and odd glides, more like a swallow than a hummingbird. Hovers with slow, deep wing-strokes and with tail spread, but often perches when feeding. Feeds on nectar, but also flying insects, hovering for long periods bouncing up and down. The distinctive call is a loud *'sweet.'*

White-bellied Woodstar *Chaetocercus mulsant*
8.5 cm (incl. 1.5 cm bill). 1800-3100 m. Short tail. Inhabits humid montane and pre-montane forest and forest edge, clearings, semi-humid bushy slopes, pasture and gardens. Has a bumblebee-like flight and forages at all levels on small flowers and insects. Submissive to all other hummingbirds but often escapes attention of territory holders by its slow insect-like flight. Attends feeders at the Machu Picchu Pueblo Hotel.

Swallow-tailed Hummingbird *Eupetonema macroura*
15-17 cm (incl. 2.2 cm bill). 1800 m. Only found at the extreme edge of the MPHS at the confluence of the Aobamba and Urubamba rivers. Usually at forest edge and open areas. A big, dark hummingbird with a longish, deeply forked blue tail (longer in male). Apparently colonizing the upper Urubamba Valley in response to deforestation.

White-bellied Hummingbird *Amazilia chionogaster*
11 cm (incl. 2.5 cm bill). 1800-3300 m. Upper mandible black, lower reddish with a dark tip. VERY similar to the next species. Tail dull bronzy-green and with more or less white inner-webs from base to tip on outer feathers – this feature not present in Green and White Hummingbird. Generally inhabits drier habitat than Green and White, but there is overlap, especially in the Sacred Valley of the Incas. Usually found in bushy terrain with cactus and agaves, *Eucalyptus* groves, bushy ravines and streamsides. Avoids moist forest. Feeds on nectar of many flowering plants and will take small insects. The call is a sharp *'zwit.'* Can be seen in the Cusichaca Valley at the start of the Inca Trail and throughout the Sacred Valley and Cusco area.

Green and White Hummingbird *Amazillia viridicauda*
Peruvian Endemic. 11 cm (incl. 2.7 cm bill). 1800-2800 m. Tail dull grayish green above and below, sometimes tipped white. Found in humid, forest-edge, secondary growth, overgrown landslides and gardens. Likes *Inga* sp. trees, but feeds on nectar of various flowers. Also hawks for small insects. Sings from an exposed perch – a series of fast squeaking notes *'tsi-tsziu-tziu twii,'* etc. Common along the Urubamba River at Machu Picchu, particularly on the grounds of the El Pueblo Hotel in Aguas Calientes where it attends feeders. Overlaps with the White-bellied in the Sacred Valley of the Incas. Study badly needed in overlap zone and perhaps only one species involved.

TROGONS AND QUETZALS   Trogonidae

Related to kingfishers, trogons and quetzals are easily recognized by their brilliant colors (bright green and red in the species below). The under-tail has distinct black and white patterns important for field identification. They are solitary forest birds sitting motionless for long periods on horizontal branches, occasionally hawking out for a passing large insect. They feed on fruits and berries as well, often hover gleaning in the canopy. They nest in holes in trees or old termite nests.

Golden-headed Quetzal   *Pharomachrus auriceps*
33 cm. 1800-3000 m. Golden-headed is somewhat similar to the Crested. Note black undertail. Found in humid montane and pre-montane forest and forest edge, generally at higher elevations than the Crested Quetzal, but there is much overlap. Encountered alone or in pairs in the mid-strata and sub-canopy of trees. In fruiting trees, hover gleans to pluck fruits. When not feeding perches motionless and can be difficult to see. The frequently heard call is a mellow *'wi-dwyyi'* repeated at regular intervals. Can be seen along the Inca Trail between Wiñay Wayna and Intipunku.

Crested Quetzal   *Pharomachrus antisianus*
32 cm. 1000-2300 m. Short crest on forehead projects over the bill. Note white undertail. Found in thick epiphyte and moss laden humid pre-montane forest. Perches quietly in the canopy and attends fruiting trees. The distinct call is a rolling *'way-way-wayoo'* and *'wheeoo.'* Can be seen near Puente Ruinas railway station.

Masked Trogon   *Trogon personatus*
25 cm. < 3300 m. The only trogon in our area. Found alone or in pairs in humid-montane and pre-montane forest in broken canopy. Sits quietly for long periods in the mid-canopy, then flies fairly long distances to a new perch. Hover gleans fruit and sallies forth for insects. Sometimes joins mixed feeding flocks. Song is a far carrying *'zooorh-hr-hr'* repeated at regular intervals. Can be seen along the Inca Trail between Phuyupatamarca and Wiñay Wayna.

MOMOTS   Momotidae

Mostly a lowland family, found only in the tropical Americas. Characterized by their long, often racket tipped tails which they frequently swing from side to side. Motmots are sluggish inconspicuous birds of forest and forest edge. They are omnivorous, feeding on a variety of insects, small mammals and fruits. They nest in burrows in banks and lay two white eggs.

Andean Motmot   *Momotus aequatorialis*
41 cm. 1800-2400 m. See Stiles (2009) for rationale for recognizing five species, four of which occur in South America. Long (25 cm) tail feathers narrow before tail tip, and become bare-shafted after wear and much preening, so that terminal rackets are separated from the rest of the tail. Found in humid pre-montane forest, mostly at the edge and in glades, also in lighter woodland and gardens. Usually encountered alone or in pairs. Sits quietly at mid-elevations, often swinging its tail from side to side like a pendulum. Sallies quickly from perch to foliage, branches or the ground to pounce on prey. The call is a soft tremulous *'hroooo.'* Quite common along the Urubamba River near Aguas Calientes and on the grounds of the Machu Picchu Pueblo Hotel.

Rufous Motmot   *Baryphthengus martii*
43-46 cm. Vag 1800- 2000 m. A lowland species recorded and photographed in the grounds of the Machu Picchu Pueblo Hotel at 2000 m on May 5, 2010. Likes tree falls, leafy gardens and forest edge. The call is a *'boop-oop'* similar to the above species but does not change pitch. Normally recorded only up to 1600 m.

PUFFBIRDS Bucconidae

Puffbirds belong to a strictly neotropical family. They are heavily built with short necks, large heads and loose plumage. They are lethargic, quiet and unobtrusive. They feed on large invertebrates. The nest is in an old termite nest or burrow in the ground. They lay white eggs.

Black-streaked Puffbird *Malacoptila fulvogularis*
19 cm. < 2100 m. Inhabits humid pre-montane forest and edge where it sits quietly in pairs, motionless for long periods, at mid to low levels. Sallies forth, after long intervals of sitting quietly, to a branch or foliage to snatch prey, mostly large invertebrates. The call is high pitched fruiteater-like *'sweeeiiii'* repeated at two-second intervals. Found mostly below 2100 m and only likely to be encountered in the lowest parts of the MPHS. Not uncommon near Puente Ruinas and around the museum on the way from Aguas Calientes to Machu Picchu.

BARBETS Capitonidae

A pantropical family with more species in Africa and Asia than in the Americas. They are thick set birds with short necks and legs and heavy bills. Brightly colored. They feed on fruit and insects and nest in tree cavities, laying white eggs.

Versicolored Barbet *Eubucco versicolor*
7 cm. < 2100. Multi-colored as its name implies. Found in humid pre-montane forest and forest edge. Usually encountered alone or in pairs at medium heights, accompanying or away from mixed feeding flocks. Perches quietly, but actively forages for fruit and rummages inside bromeliads. Found mostly below 2000 m and only likely to be found at lower elevations within the MPHS.

TOUCANS Ramphastidae

Mostly tropical though several species range up and in to humid montane and pre-montane forest. Easily recognized by enormous, colorful bills which are surprisingly light. They are usually found in pairs or small groups. When a group decides to fly, one will fly first followed by others, one by one. Toucans fly directly with fast wing beats, often gliding between each beat. Mostly frugivorous, they will sometimes eat other things including nestlings and eggs of other birds. They nest in natural or woodpecker holes in trees.

Black-throated (Emerald) Toucanet *Aulacorhynchus (prasinus) atrogularis*
34 cm. < 3000 m. Considered by some as part of Emerald Toucanet *A. prasinus*, but see Puebla-Olivares *et al.* (2008) who identified three clades in South America based on DNA and proposed species rank for each. Note the black bill with yellow upper ridge. Note black throat. Mainly a bird of humid pre-montane forest, but will wander to nearby man made clearings and lighter woodland. Typically in pairs or small family groups. Most often seen eating small fruits. The call is a series of low pitched *'churt'* notes. This small green toucan is rare at Machu Picchu and only likely to be seen at the lowest elevations, below 2100 m. Best looked for in the lower Mandor and Aobamba Valleys.

Blue-banded Toucanet *Aulacorhynchus coeruleicinctus*
40 cm. 1800-2500 m. Note white throat and blue band across chest. Behavior and habits similar to previous species, but the call is a rather different *"kirrit-it kirrit ik ik"* etc. More confiding and generally replaces above species at higher elevations. Quite common near Machu Picchu.

Grey-breasted Mountain Toucan *Andigena hypoglauca*
48 cm. 2000-3500 m. Large with colorful bill. Rump yellow, very prominent in flight, contrasting with the black tail. Inhabits humid montane and pre-montane forest and wooded ravines. Omnivorous. Found in pairs or small family groups. The call are a nasal *'kuaaaaaa'* and a sharp *'kip-kip-kip-kip.'* Encountered most frequently along the Inca Trail between Phuyupatamarca and Intipunku.

WOODPECKERS Picidae

A cosmopolitan family found worldwide except in Australia. A few species are adapted to a treeless environment, but most are recognized by their habit of excavating holes in trees and "drumming" on resonant parts of trees as part of territorial advertisement. Most have stiff tails adapted for support to aid in the climbing of trees. Invertebrates are found by excavating and tapping on decaying wood. Some seeds and berries are also eaten. Woodpeckers fly with a characteristic undulating flight. They nest in excavated holes in trees, some (flickers) excavate holes in banks and old adobe buildings.

Ocellated Piculet   *Picumnus dorbygnianus*
10 cm. < 2500 m. Ocellated = eyelike markings. The only piculet in the Machu Picchu area. Behaves like a nuthatch, hanging upside down on branches looking for invertebrates. Usually seen singly or in pairs, often accompanying mixed species flocks, in humid montane and sub-montane forest with lots of epiphytes. The song is typical of the genus – a high pitched trill. Quite common along the road and railroad track between Aguas Calientes and the Mandor Valley.

Bar-bellied Woodpecker   *Veniliornis nigriceps*
19 cm. 2400-3500 m. A small woodpecker, heavily barred below with buff and blackish. Seen singly or in pairs, sometimes with mixed species flocks, in humid montane and elfin forest with an understory of scrub and particularly bamboo (*Chusquea* sp.) for which it shows a preference. The call is a high pitched, descending *'kzzrr.'* Fairly common throughout the humid forests of the MPHS and at Abra Malaga.

Golden-olive Woodpecker   *Piculus rubiginosus*
23 cm. < 2400 m. A medium sized woodpecker. Note white cheek and heavy barring on the undersides. Usually single birds are seen, often accompanying mixed feeding flocks, in the mid-story or canopy of humid montane and sub-montane forest and borders. Feeds by chiseling and hammering on limbs and vines. Also, probes in mossy growths. Makes a variety of harsh calls. Most commonly encountered in heavy, tall forest. Can be seen near along the railroad track at Machu Picchu at Aguas Calientes.

Crimson-mantled Woodpecker   *Piculus rivolii*
28 cm. < 3500 m. The *atriceps* race is present at Machu Picchu. Another medium sized woodpecker, unmistakable with its almost all crimson upper-parts. Seen singly or in pairs, often accompanying mixed feeding flocks. Feeds by probing and searching rather than hammering, at all levels of the forest and sometimes on the ground. A bird of montane and pre-montane forest that can be found wherever there are trees within the Sanctuary, sometimes right up to tree-line at 3500 m. Can be seen between Phuyupatamarca and Wiñay Wayna along the Inca Trail and at Abra Malaga. Named for the French ornithologist and collector François Massena Prince d´Essling, Duc de Rivoli (1799-1863).

Andean Flicker   *Colaptes rupicola*
30 cm. 2500-4500 m. Striking, mostly terrestrial woodpecker, with a special place in local mythology. Prominent bright golden rump in flight. Quite gregarious, loose flocks move across open country, mostly among rocks and small cliff faces, but will use scattered trees if present, probing into crevices and cracks and digging into the ground with its long beak. Very vocal. Calls include *'chew-chew-chew' 'kuaa-ap-kuua-ap'* and a loud *'keek.'* Nests in holes in banks and old adobe buildings. A bird of the high puna grasslands, avoids wooded areas. Often encountered by hikers in the high country and a common sight along the higher parts of the Inca Trail, the Sacred Valley and Cusco areas, seldom descending below 2500 m in our area.

Crimson-bellied Woodpecker   *Campephilus haematogaster*
33 cm. < 2300 m. Large striking, mostly black woodpecker with a crimson belly, rump and lower back. The only member of this genus to occur at Machu Picchu. Found alone or in pairs in mature humid sub-montane forest, often quite low down. Searches for invertebrates on large tree trunks, hammering and probing in crevices. Rare in the MPHS, likely only to be encountered in the lowest, more isolated reaches of the MPHS, in the Mandor and Aobamba Valleys.

PARROTS Psittacidae

Parrots inhabit the Americas, Africa, Asia and Australia with the greatest diversity in the tropics. They are easily recognized by their noisy social habits and heavy hooked beaks. They range in size from sparrow-sized parrotlets to large macaws. Most common in the Amazonian lowlands, many species range to the Andes. Sexes are alike and mating is for life; pair bonding is maintained by mutual preening and play behavior. Parrots are arboreal and feed on fruits, seeds, nuts, flowers and blossoms, often raiding crops in the Andes where some species are considered pests. They are mostly sedentary, moving locally between feeding sites. Parrots nest usually in tree holes or holes in cliffs and lay between 2-5 white eggs. The female incubates the eggs, but both parents feed the young. Recent comprehensive genetic analyses (Hackett *et al.* 2008) indicate that the closest relative is most likely the Passeriformes or the Falconiformes, as also recently found by Suh (2011).

Golden-plumed Parakeet *Leptosittaca branickii*
34 cm. 2700-3350 m. Endangered. Inhabits humid montane and sub-montane forest. Social and usually encountered in noisy groups of 6-15 feeding in the canopy Also in small bushes on mountain slopes and often flying high over passes. The harsh call is a rather macaw-like *'rhaaa-aa.'* Rare. Named for Wladyslaw Graf von Branicki (1864-1926), of the Branicki Zoological Museum Warsaw.

Mitred Parakeet *Psittacara mitrata*
38 cm. < 4300 m. Two races currently recognized, may involve two species, both of which occur in the Cusco area. The nominate *mitrata* is a bird of humid forests found in large noisy flocks feeding in tree canopies, often in flowering *Erythrina* sp. trees in the Urubamba Gorge at Aguas Calientes and Puente Ruinas. Undersides of flight and tail feathers old gold in color. *A. m. alticola* found at higher elevations in more arid temperate zone near Llactapata and the Sacred Valley of the Incas often raiding maize crops and other cultivations. Calls include a deep harsh *'cherree'* and a snarling *'whee-eee-rhee.'* Limits of distribution and habitat in these two forms not fully understood.

Barred Parakeet *Bolborhynchus lineola*
17 cm, with a 5-6 cm pointed tail. < 3300 m. Inhabits humid montane and pre-montane forest, forest edge and clearings. Seems to particularly like seeding bamboo (*Chusquea* sp.). Inconspicuous and mostly noted as flocks of 5-20 fly directly over the treetops, sometimes very high. Feeds sluggishly on bamboo, *Cecropia* catkins, buds, seeds and flowers in the mid to understory. Difficult to detect when perched. Very erratic in habits and semi-nomadic, depending on the flowering cycle of the bamboo.

Andean Parakeet *Bolborhynchus orbygnesius*
16 cm, with 6 cm pointed tail. 2400-3900 m. Tail broad at the base. Outer webs of primaries blue-green. Mostly found in semi-arid situations, bushy slopes and ravines, dry cloud forest etc. Usually encountered in groups of 5-30, feeding in bushes, brambles and small trees or on the ground for seeds, berries and fruits. Raids corn crops. Flight is swift and direct. Call notes includes a chattering *'rueet-rueet-rueeet'* also a twittering *'dydydee-dee-dee.'* Can be seen in the bushy scrub between Paucarcancha and Pampacahuana along the Inca Trail and at Abra Malaga.

Speckle-faced Parrot *Pionus tumultuosus*
29 cm. 1800-3300 m. Mostly green with a speckled, plum colored head. Inhabits humid montane and sub-montane forest where it is primarily found in treetops feeding quietly in pairs or small flocks flying across valleys. Will raid maize crops. Flies with deep wingbeats typical of the genus.

Scaly-naped Amazon *Amazona mercenaria*
33 cm. 1800-3400 m. A large stocky parrot with broad blunt wings. Wings purplish-blue with a red wing speculum. Found in humid montane and sub-montane forest with tall trees, mostly along well forested ridges. Singles, pairs, small and large groups can be encountered. Often in large flocks when moving from one feeding area to another. Usually seen flying high with characteristically stiff, shallow and fluttering wingbeats while calling constantly. Feeds in treetops, very shy when perched. The call is a loud harsh *'kree-rrhhee'* or *'khoeee-khooeouu.'* Can be seen at Abra Malaga.

Golden-plumed Parakeet

Mitred Parakeet

Barred Parakeet

Andean Parakeet

Speckle-faced Parrot

Scaly-naped Amazon

ANTBIRDS Thamnophilidae

A large diverse family, strictly neotropical, attaining its highest diversity in the lowland Amazonian rainforest. Species range from large antshrikes to small warbler-sized antwrens. Monogamous, pairs keep in close contact by song and calls. All are strongly sexually dimorphic: males being shades of gray and females shades of brown. They live in shady vegetation and feed on invertebrates. The nest is an open cup built above the ground. Both sexes build the nest, brood and tend the young.

Variable Antshrike   *Thamnophilus caerulescens*
14.5 cm. < 2400 m. The only antshrike at Machu Picchu where the distinctive *melanochrous* race is present. Encountered at mid-levels (2-5 m) in thickets and undergrowth of pre-montane forest and forest edge, often in more open woodland. For an antbird, relatively easy to see and often confiding. Invariably found in pairs with, or away from, mixed species flocks. Quivers tail occasionally and often raises and lowers it. Feeds by working its way up vine tangles and thick bushes, working out to branch ends, looking for caterpillars, spiders and other invertebrates. The song is a distinctive *'cow-cow-cow-cow-cow-cow,'* also a single call note *'caw.'* Quite common between Aguas Calientes and the Mandor Valley at Machu Picchu, this being one of the best localities in Peru to observe this species.

Streak-headed Antbird   *Drymophila striaticeps*
15 cm. < 2500 m. Recently split form of the Long-tailed Antbird *(Drymophila caudata)* Isler *et al.* (2012). Distinctive with long tail and streaked plumage. Occurs in dense stands of *Chusquea* sp. bamboo to which it is tied. Pairs forage in the crowns of these bamboo thickets and vine tangles within. Often with mixed flocks but not always. Song is a buzzy, sliding *'chip-chip chew-chew-chew-chew'* often duet ting with higher pitched female. Can be seen on the slopes of Machu Picchu in bamboo.

ANTPITTAS Grallaridae

Antpittas are subdued in color, plump with short tails and long legs, being mostly shades of brown and behaving like rails on land. The smaller antpitttas *(Grallaricula)* are more arboreal, inhabiting dense thickets, usually within 3 m of the ground. Obviously similar to their old world counterparts – pittas *(Pittidae)* – in shape if not in color. Mostly solitary. They walk and hop on or near the ground. Generally shy and difficult to see, their simple, far-carrying songs betray their presence.

Undulated Antpitta   *Grallaria squamigera*
20-22 cm. 2000-3500 m. A very large unmistakable antpitta. Sexes similar. Found alone or in pairs on or close to the ground in thick vegetation in montane and elfin forest. Sometimes hops into adjacent grassy clearings, particularly at dawn and dusk. Shy and retiring, feeds mostly on the ground, but will hop quite high in bushes and trees, often freezing on a branch when alarmed. The song, usually given from an elevated perch, is a series of quavering notes, quite haunting in quality, lasting 4-5 seconds and repeated every 5-10 seconds or so – *'hohohohohoho-ho-ho-ho-ho.'* Can be seen near Sayacmarca on the Inca Trail and at tree line at Abra Malaga.

Scaled Antpitta   *Grallaria guatamalensis*
16-16.5 cm. < 2000. Inhabits the understory of humid pre-montane forest and well developed secondary woodland. Terrestrial and secretive, often in tangled vegetation alongside streams and in ravines. The song is a series of low-pitched, hollow resonant notes which slide up in pitch and become louder, lasting 4-5 seconds. Found at elevations from lowlands to about 2000 m. At Machu Picchu this species is at the limit of its elevational range, but has been recorded near Puente Ruinas railway station.

Variable Antshrike

Streak-headed Antbird

Undulated Antpitta

Scaled Antpitta

Stripe-headed Antpitta *Grallaria andicolus*
16-16.5 cm. 3500-4600 m. The *punenis* race, which may deserve full species rank as Puna Antpitta, is present in our region. Krabbe & Schulenberg (2003) noted that vocal differences suggest that *punensis* should be treated as a separate species. Found alone or in pairs in elfin forest at tree line and particularly in *Polylepis* woodland patches above tree line. Forages on the ground, but sometimes hops quite high up on branches, especially when singing. Sometimes forays out to open grassy patches with boulders and rocks. The song is an often heard frog-like rolling *'reee-reeee'* and the call is a far carrying descending one or two note whistle, delivered from high up in a bush. Can be seen in isolated, small *Polylepis* patches near Llulluchapampa and Abra Warmiwañusca on the Inca Trail and is common in the same habitat at Abra Malaga and other high altitude woodlands in the Sacred Valley. Mostly above 3000 m.

Red and White Antpitta *Grallaria erythroleuca*
Peruvian Endemic. 17-17.5 cm. 2100-3000 m. Found alone or in pairs in dense thickets in humid montane and pre-montane forest, often dominated by a *Chusquea* bamboo understory. Feeds mostly on the ground but will hop up onto low branches, especially when singing. Shy and difficult to see without use of sound playback. The song is a far carrying three-note whistle – the last note slightly descending – *'tew-hoo-hoo'* rarely a fourth note is added. The call is an explosive *'tuew.'* Fairly common at Abra Malaga near San Luis.

Rufous Antpitta *Grallaria rufula occabambae*
14.5-15 cm. 2400-3700 m. The *occabambae* race is present at Machu Picchu. Found, like all its congeners, alone or in pairs, mostly in dense foliage just above the ground in montane and elfin forest and forest borders, favoring mossy, boggy areas and streamsides, showing a preference for a *Chusquea* bamboo understory. Difficult to see. The song in our region is a two-noted whistle on the same pitch, the first note stronger – *'pee-pee.'* Geographical variation in song throughout the range of this species is strong and taxonomic work is revealing that several distinct species are involved. Found mostly at higher elevations than the preceding species. Common at Abra Malaga.

Rusty-breasted Antpitta *Grallaricula ferrugineipectus*
11-12 cm. 2000-3250 m. The *leymebambae* subspecies is present in our region. Distinctive triangular white patch behind eye. Found usually in pairs and very difficult to see. Inhabits fairly open understory of humid montane forest within a meter of the ground, feeding by hopping along moss covered stems and branches. Flicks wings and tail, sometimes fly-catches. Song is a high pitched *'kee-ke-kee-kee-kee-kee-kee-ki.'* Krabbe & Schulenberg (2003) indicated that vocal differences suggest that the southern subspecies *leymebambae* deserves recognition as a separate species. Uncommon and rarely seen in our region.

TAPACULOS Rhinocryptidae

A mainly Andean family of small birds that live in thick vegetation and usually cock their tails upright. Almost flightless, they live mostly on the ground. Seldom seen, their loud songs are often heard. They build a round nest with a side entrance that is placed low in a bush, grass tussock or at the end of a tunnel. The taxonomic situation of this family is complex and in flux, with new species still being described. Very difficult to see and/or identify except by song or phylogenetic analyses.

Trilling Tapaculo *Scytalopus parvirostris*
11 cm. 2100-3600 m. Very difficult to see (without the use of sound playback), behaves more like a mouse than a bird, creeping around in the dense undergrowth of humid montane and pre-montane forest, alone or in pairs. Forages mostly on the ground in the densest, darkest vegetation. The song is a long drawn out trill of 12-16 notes lasting about 18 seconds. Can be heard, though seldom seen, along the Inca Trail between Wiñay Wayña and Machu Picchu and at Abra Malaga.

Diademed Tapaculo *Scytalopus schulenbergi*
11 cm. 2750-3400 m. Diadem = headband. This species inhabits the upper humid montane forest with much bamboo (*Chusquea* sp.), between the elevational ranges of *S. simonsi* (higher) and *S. parvirostris* (lower), but only on the right bank of the Urubamba River near the Wakay Wilca massif. The song is a rapid trill-like *S. parvirostris*, but slightly deeper pitched. Fairly common at Abra Malaga. Named for Thomas S. Schulenberg (b. 1954) US ornithologist and lead author of *Birds of Peru* (2010).

Vilcabamba Tapaculo *Scytalopus urubambae*
Peruvian Endemic. 9.5-10 cm. 3500-4200 m. Inhabits humid montane and elfin forest, often in areas dominated by moss and boulders. Hops and creeps around in nooks and crannies in boulder fields and dense mossy vegetation at tree line. More confiding than other members of the genus. Song is a raspy *'chit-chit-chit'* repeated for long periods. Can be readily seen around the Salkantay massif and along the Inca Trail between Sayacmarca and Phuyupatamarca. Replaces the following on the left bank of the Urubamba River.

Puna Tapaculo *Scytalopus simonsi*
10 cm. 2900-4300 m. Inhabits high habitats at tree line, rocky and bushy ravines and hill-slopes with screes and boulders. Also, partial to patches of *Polylepis* woodland. Hops and creeps around in nooks and crannies in boulder fields, and dense mossy vegetation, sometimes hopping up on branches of *Polylepis* trees and boulders, from which it will sing. Sometimes quite tame and easy to see. The song is a repeated *'kirk-kirk-kirk-kirk'* often delivered for long periods. Replaces Vilcabamba Tapaculo (*S. urubambae*) on the east side of the Urubamba River and only likely to be encountered near tree line on or near the Wakay-Wilca massif. Common in *Polylepis* at Abra Malaga.

ANTTHRUSHES Formicariidae

Antthrushes are long-legged secretive antbirds that walk deliberately across the forest floor with cocked tails. Much more often heard than seen.

Rufous-breasted Antthrush *Formicarius rufipectus*
18 cm. < 1900 The *thoracicus* race is present in the MPHS. Inhabits humid pre-montane forest with an understory of saplings and fallen trees, also thick mature secondary growth. Totally terrestrial and runs around on the ground like a rail. The song is a two note loud ringing whistle delivered in one second, repeated about every 10 seconds. Difficult to see and usually only detected by its song. Only likely to be encountered in the lower parts of the Sanctuary such as the lower Aobamba Valley.

Barred Antthrush *Chamaeza mollissima*
19 cm. 1800-3100 m. The *yungae* race is at Machu Picchu. Found in the dense understory of humid montane and pre-montane forest where there are fallen trees covered in moss. Very hard to see and completely terrestrial running like a rail, pumping its cocked tail. Sings from a slightly elevated perch such as a log. The song is a very long trill rising in volume and accelerating for 20 seconds, ending abruptly.

OVENBIRDS (including WOODCREEPERS) Furnariidae

A large and diverse family, found exclusively in the neotropics reaching its highest diversity in the Andes of Peru. Ovenbirds occupy almost all habitats from lowland Amazonian rainforest up to snow line. Now usually deemed to include the woodcreepers. Most are a shade of brown with short rounded tails and primitive tail feathers reflecting their poor flying abilities. Sexes are similar and they are monogamous, keeping in close pairs. Both sexes sing, build nests and take care of fledglings. Most species have a large rounded nest structures with a side entrance. Woodcreepers forage by hitching up tree trunks and along branches supported in part by stiffened tips to the tail feathers. They nest in cavities in trees. This classification treats woodcreepers (formerly *Dendrocolaptidae*) and the ovenbirds *(Furnariidae)* as members of a single family. Whether the two groups are sister taxa has never seriously been questioned (see Sibley & Ahlquist 1990, Marantz *et al.* 2003, Remsen 2003). Historically, the controversy centered around the taxonomic ranking of the two groups, with some authors treating them as subfamilies of the same family, whereas others treated each as separate families.

Slender-billed Miner *Geositta tenuirostris*
18-19 cm. > 2700 m. Long, slender, slightly decurved bill distinguishes this bird from other miners and its short tail from the earthcreepers. In flight shows lots of rufous in the wing and tail. Found on puna grasslands and meadows, often feeding in adjacent plowed fields and corrals, alone or in pairs, on the ground. Walks with a waddling gait. Males fly high during display, uttering a high pitched, repetitive *'tji-tji-tji-tji.'* When flushed gives a surprised *'eeek.'* Fairly Common.

Common Miner *Geositta cunicularia*
15-17 cm. > 3500 m. An inconspicuous bird of open country, this species crouches to avoid detection but flies long distances when flushed. Mostly brown with creamy rump and rufous wing panel obvious in flight. Most often found in rolling open country, preferring some cover such as rocks or small bushes. True to their name, they excavate and nest in tunnels in banks, but contrary to their name they are not common in our area. Can be seen in higher areas of the MPHS, the Sacred Valley and Abra Malaga.

Strong-billed Woodcreeper *Xiphocolaptes promeropirhynchus*
28-30 cm. < 3100 m. A large robust woodcreeper with a heavy, slightly decurved bill. The highland race found at Machu Picchu *(lineatocephalus)* may be a separate species from the lowland Amazonian race. Easily identifiable by its large size alone. Found in mature sub-montane forest at all levels often with mixed feeding flocks. Feeds by climbing large tree trunks searching for invertebrates. The song is far carrying series of descending whistled notes – *'pt-tew pt-tew pt-tew pt-tew'* most often heard at dawn. Rare in the MPHS.

Greater Scythebill *Drymotoxeres pucherani*
28 cm 2100-3650 m. Note huge decurved bill and dark cheek patch. Little known and usually found alone, with or without flocks. Probes in moss and especially bromeliads. The call is a rising whining note and a descending series of whistles *'eek-eek pee-eek pee-rwee reew.'* Has been seen on higher sections of the Inca Trail.

Montane Woodcreeper *Lepidocolaptes lacrymiger*
19 cm. < 3200 m. A medium sized woodcreeper with a longish decurved bill typical of the genus. Found in mature humid montane and sub-montane forest and nearby lightly wooded clearings. Feeds in typical woodcreeper fashion, hitching up trunks and undersides of limbs, searching cracks and crevices for insects. More often than not, seen with mixed feeding flocks. The seldom heard song is a series of thin whistled notes. Not uncommon in the Sanctuary, but rarely seen. Can be seen on the Inca Trail between Wiñay Wayna and Machu Picchu.

Streaked Xenops   *Xenops rutilans*
12 cm. < 2100 m. The only xenops at Machu Picchu. A small arboreal furnariid. Found in humid pre-montane forest and forest edge as well as second-growth and lighter woodland. Forages in the mid to lower story by working along or underneath slender branches like a piculet, swiveling from side to side, flaking off dead wood or bark. Often with mixed species flocks. The call is a thin series of descending notes. Only likely to be seen in the lower parts of the MPHS. Not uncommon along the railway track near Puente Ruinas railway station.

Streaked Tuftedcheek   *Pseudocolaptes boissonneautii*
20-21 cm. < 3450 m. A large distinct furnariid. Note prominent white cheek tufts. Usually encountered singly within mixed species flocks and invariably investigating the centers of epiphytes for invertebrates, where they often stay for long periods, hammering and rummaging. The call is a harsh, sharp *'chit'* and the song is a combination of two *'Chit'* notes followed by a fast trill. Usually well above the ground in humid montane and elfin forest. Can be seen throughout the MPHS and at Abra Malaga in suitable habitat.

Rusty-winged Barbtail   *Premnornis guttuliger*
13 cm. < 2500 m. Unobtrusive in humid pre-montane forest understory. Does not creep along branches and is often found with mixed understory flocks, looking like a small foliage-gleaner, searching among leaves, often dead leaves. Streaked with rufous tail and wings. Can be looked for along the railway track between Puente Ruinas and the Aobamba Valley.

Sharp-tailed Streamcreeper   *Lochmias nematura*
15-16 cm < 2300 m. The *obscurator* race is present. Almost always found along forested mountain streams near or on ground where it hops along the margins or on rocks probing and flicking leaves as it goes. Inconspicuous. The call is rising musical chatter *'tur-tr-tr-tr-tr ti TI-TI TI-TI.'* Also a buzzy rattle. Found on streams in the vicinity of Aguas Calientes and undoubtedly suitable streams lower in the MPHS.

Wren-like Rushbird   *Phleocryptes melanops*
14 cm 3200-4200 m. Found exclusively in beds of medium to tall reeds, especially *schirpus*. Moves between patches of reeds with short buzzy flight. Call is series of non-birdlike *'tic'* notes like tapping two pebbles together. Common at Huacarpay Lakes south of Cusco and on lakes and ponds in the Chinchero area.

Buff-breasted Earthcreeper   *Upucerthia validirostris*
19 cm. 3400-4800 m. Although the *jelskii* subspecies group present in our area has been considered a separate species from *U. validirostris*, genetics now show this to be invalid. Therefore, we return to earlier classification as recommended by the South American Classification Committee. Long decurveddecurved bill and buff breast. Tail often half-cocked as the bird (usually alone) runs and hops across the ground. Poor flyer, preferring to hide behind grass tussocks or rocks to avoid detection. Excavates holes in banks to nest. The song is delivered from the top of a rock or stone wall and consists of a weak trill. The call is a harsh *'chick.'* They feed by probing their sickle-shaped bill into loose, soft soil. Found mostly in open grassland with rocks and banks. Fairly common at high elevations.

Montane Foliage-gleaner   *Anabacerthia striaticollis*
16 cm. < 2100 m. The only foliage-gleaner found within the Sanctuary. A conspicuous whitish eye-ring and narrow post-ocular streak, gives the bird a "spectacled" appearance. A fairly easily observed bird of sub-montane forest and forest borders and edge. Forages in an active manner at mid and upper levels, often hanging upside down as it probes into dead leaf clusters, on which it appears to specialize. Often accompanies mixed species flocks, but just as likely to be encountered alone or in pairs. The call is a sharp *'pick'* note and the song is a nondescript dry series of notes given at dawn. Only likely to be encountered in the lower parts of the MPHS such as in the Mandor and Aobamba Valleys.

Cream-winged Cinclodes *Cinclodes albiventris*
16-17 cm. > 2750 m. A fairly common bird. Distinctive creamy off-white wing band shows in flight. The cinclodes to learn well so as to compare it with the following two species. Usually encountered alone or in pairs. Highly territorial and often seen chasing each other with much tail-cocking and wing-raising. Sings from the top of a rock while wing-raising – a short fast trill, one of the most familiar bird songs of the high country. Inhabits open grassy areas, not exclusively along streams or wet bogs, often in quite bushy country and farmlands. A familiar bird within the MPHS and at Abra Malaga.

Royal Cinclodes *Cinclodes aricomae*
Peruvian Endemic. 20 cm. 3700-4600 m. Critically endangered. Large with mottled breast and contrasting white throat. Not likely to be confused with other cinclodes species in the area as its habitat differs markedly. Found exclusively in patches of *Polylepis* woodland on steep, rocky slopes. Seems to need mossy rocks and branches to feed, where it flakes off large pieces of moss, searching for insects and grubs. Sings from a branch or top of a bush. The song is a very loud trill. Can be seen in *Polylepis* groves at Abra Malaga, Mantanayoc above Yanahuara and other *Polylepis* groves in the Cordillera Urubamba. A rare, globally threatened species. Named for the type locality Aricoma pass in Puno (no *Polylepis* left)

White-winged Cinclodes *Cinclodes atacamensis*
19-21 cm. > 2800 m. Larger than the preceding species with extensive white wing patch and band on the primary coverts especially prominent in flight, but easily visible on the closed wing. More rusty upperparts. Behavior is similar to the preceding species. The song similar if somewhat louder. This species is restricted to clear running streams, almost never being found away from them. Mostly encountered above 3000 m up to snow line. Can be seen at Sisaypampa above Pampacahuana and along the Urubamba River.

Rufous-backed (Peruvian) Treehunter *Thripadectes scrutator*
22-24 cm. 2100 -3500 m. Hard to see and seldom encountered. Always alone, away from mixed flocks, foraging in the undergrowth of humid montane forest, often where there are large stands of *Chusquea* sp. bamboo. Birds move slowly, working their way along stems, often flicking their wings. The song is a loud descending rattle which accelerates toward the end. Rare or overlooked. Can be seen at Abra Malaga.

Striped Treehunter *Thripadectes holostictus*
20-21 cm. < 2400 m. Very similar to the preceding species but usually found at lower elevations. Equally difficult to see. Differs from Buff-throated Treehunter by streaking on upper-parts extending down onto lower back. Behavior and habitat similar to that species, but seems more likely to be found with mixed species flocks. The song is a fast, slightly descending trill. Likely to be encountered in dense forest.

Spotted Barbtail *Premnoplex brunnescens*
13-14 cm. < 2500 m. A smallish, obscure, dark brown furnariid. Can be found alone or with mixed species flocks. Climbs trunks and branches in dark forest undergrowth. The infrequently heard song is a fast high pitched trill. Inhabits humid montane and sub-montane forest. Best looked for in the Mandor and Aobamba Valleys.

Pearled Treerunner *Margarornis squamiger*
14-15 cm. 2100-3700 m. Usually found with mixed species flocks in groups of 1-8 individuals, one of the more conspicuous members of these flocks. The infrequently heard song is a single weak trill or spaced single notes. Hitches along mossy trunks and limbs in wet montane and elfin forest at mid-levels. A rather common bird of the humid temperate forest, often seen along the Inca Trail between Sayacmarca and Phuyupatamarca and at Abra Malaga.

Tawny Tit-Spinetail *Leptasthenura yanacencis*
17 cm. 3900 -4500 m. Cinnamon colored all over. Restricted to isolated patches of *Polylepis* woodland where it is encountered in pairs or small family groups. Typically feeds from inside a bush, working its way to the outermost tips of the branches. The song is often heard, consisting of a variety of trills and chatters. A rare species, found off the beaten track in isolated *Polylepis* woodlands. Can be seen in *Polylepis* groves at Abra Malaga, Mantanayoc above Yanahuara and other *Polylepis* groves in the Cordillera Urubamba.

White-browed Tit-Spinetail *Leptasthenura xenothorax*
Peruvian Endemic. 15 cm. 3800-4500 m. Endangered. Quite different from the above species. Note rufous cap and clear belly. Usually found in small groups feeding on larger limbs of trees. The song is a high pitched two second trill – *'trrrrrrrrrrr.'* Contact call is a sharp *'chit.'* Another *Polylepis* woodland specialist. Can be seen in *Polylepis* groves and adjacent *Gynoxys* thickets at Abra Malaga, Mantanayoc above Yanahuara and other *Polylepis* groves in the Cordillera Urubamba. A rare, globally threatened species.

Andean Tit-Spinetail *Leptasthenura andicola*
16 CM. 3500-4200 m. Whitish superciliary and heavily streaked back. In our region this species does not occupy *Polylepis* woodlands where the White-browed Tit-Spinetail *Leptasthenura xenothorax* is found. Encountered in low brush and scrub. Rare.

Streak-fronted Thornbird *Phacellodomus striaticeps*
17 cm. 2900-4150 m. Usually in pairs or small family groups, skulking and foraging low or on the ground. Prefers thorny scrub with cacti in our area. Tail may be partly cocked. Song is a loud trill and cascade given in duet *'kyew-kyew-kyew-kyew KYEW-KYEW-KYE.'* Fairly common in xerophytic scrub near Huacarpay Lakes south of Cusco.

Line-fronted Canastero *Asthenes urubambensis*
16-16.5 cm. 3100-3800 m. The nominate *urubambensis* race in our region. Heavily streaked below almost unstreaked above. More arboreal than other members of the genus, hopping along branches with tail slightly elevated, but when on ground walks like a pipit probing mossy clumps. The song is a thin ascending trill. Usually solitary inhabiting *Polylepis* groves, elfin forest and mossy slopes with small bushes.

Junín Canastero *Asthenes virgata*
Peruvian Endemic. 17-17.5 cm. Peruvian endemic. Heavily streaked above and below, note the streaking on the back which the preceding species lacks. Behavior like others of the genus – individuals or pairs run quickly between grass tussocks and small bushes. The song is typical also, a descending series of buzzy notes – *'tzee-tzee-tzee-tzee-tzee-tzee-trreeeeee'* delivered from atop a small bush or tussock. Found at tree line where elfin forest or semi-humid scrub grades into grassland with small bushes and ferns, or in lush ungrazed tussock grass. Common in the grassland/*Polylepis* interface at Abra Malaga.

Scribble-tailed Canastero *Asthenes maculicauda*
17-18 cm. 3000-3900 m. Very similar to the proceeding species but note unstreaked rufous forecrown and pale throat. Central tail feathers extensively "scribbled" with rufous and black (lacking or faint in Junin Canastero). In our area this and the preceding species overlap. This species prefers more humid paramo with taller grass. Song similar to Junin Canastero – a dry *'tzreee-tzreee-tzree-ti-ti-ti-ti-ti-ti-tu'* Also a liquid *'hueeet.'* At present only known from above Mantanayoc in the Sacred Valley.

Streak-backed Canastero *Asthenes wyatti*
15.5 cm. 3500-4600 m. The race which occurs at Machu Picchu is the southern *graminicola* race which has no streaking below. Shows a strongly rufous sided tail in flight, with the central feathers being dusky brown. As with its congeners, mostly terrestrial, running between tussocks and rocks with tail half-cocked. The song is a fast ascending trill delivered from the top of a rock or bush. Rare within the MPHS and at Abra Malaga, occurring in high puna grasslands, in tussock grass (*Ichu* sp.) with some rocks or bushes. Named for English ornithologist Claude Wyatt (1842-1900) who collected in Colombia.

Streak-throated Canastero *Asthenes humilis*
15 cm. 3700-4800 m. Similar to the next species but with all brown (not rufous) tail and faint streaks on upper-parts. Does not cock its tail. Song is a repeated *'trrr-trrr trrr-trrr trrr-trrr'* delivered from a grass tussock. Also *'pit-pit-pit.'* Found alone or in pairs on high puna grasslands with grass tussocks, often near water and dry-stone walls. Common at Abra Malaga.

Cordilleran Canastero *Asthenes modesta*
14.5-15 cm. 3600-4600 m. Long, often cocked tail. No streaking on the upper-parts. Mostly terrestrial and mostly observed alone. Walks and runs with tail cocked, between rocks, bunch grass and small bushes. Sings from the top of a small bush or rock – a monotonous sharp descending trill. Found in open, somewhat arid grassy areas, with some rocks. Can be seen at Llulluchapampa along the Inca Trail and at Abra Malaga.

Rusty-fronted Canastero *Asthenes ottonis*
Peruvian Endemic. 18.5cm. 2900-4000 m. Wings and long, thin, graduated tail rufous-chestnut. Forehead rufous. Found alone or in pairs low to the ground in bushes and small trees, sometimes running across the ground or rocks with tail cocked. Not likely to be found in humid montane forest, prefers semi-arid scrub and bushes. The song is a buzzy descending *'bzee-bzee-bzee-bzee-zee-zee-dd-dd-dddddd'* delivered from the top or inside of a bush. Most easily seen in the Cusichaca Valley between the Llactapata ruins and Wayllabamba on the Inca Trail, the west side of Abra Malaga and at Huacarpay Lakes south of Cusco. Named for German Otto Garlepp who collected in Peru 1895-1912.

Puna Thistletail *Asthenes helleri*
18 cm. 2700-3700 m. Another small furnariid with a long and untidy tail. Found singly and in pairs hopping through dense vegetation with tail cocked. Does not associate with mixed feeding flocks. The call is a repeated *'preek.'* The song is a high-pitched descending trill. Fairly common in the undergrowth of humid montane and elfin forest near tree line within the MPHS and Abra Malaga. Fairly common. Named for US zoologist Edmund Heller (1875-1939), he was employed as a naturalist/collector on the 1915 expedition to explore the newly discovered lost city of Machu Picchu.

Azara's Spinetail *Synallaxis azarae*
17-18 cm. < 3100 m. Gray with a rufous cap and wings and untidy tail. Found usually in pairs, this species inhabits dense undergrowth, grass and ferns close to the ground in humid montane and premontane forest, and rarely seen higher. Difficult to see and rarely perches in the open for more than a second. Most often detected by the monotonous repeated call note – *'pip-sqeek'* given throughout the day as the bird moves undetected through dense vegetation. Found in bushy country and forest edge. Can be found along the railway track from Aguas Calientes to Puente Ruinas and near the Machu Picchu ruins. Named after Félix Manuel de Azara (1746-1811) Spanish military officer, engineer and naturalist who collected and described hundreds of birds and mammals during his years in South America (1781-1801).

Marcapata Spinetail *Cranioleuca marcapatae*
Peruvian Endemic. 16 cm. 2400-3350 m. The nominate form occurs in our region. Note rufous cap bordered by black and white eye lines. This species has only been recorded on the right bank of the Urubamba River (nominate *marcapatae* race) but further studies could reveal that the white crowned *weskei* race or intermediates occur on the left bank. Found singly and in pairs, often accompanying mixed species flocks, in lower growth and edges of humid montane and elfin forest, showing a particular liking for *Chusquea* sp. bamboo, to which it seems to be closely tied. Hops and hitches along small branches and limbs searching and probing for insects. The song is a thin descending trill. Can be seen at Abra Malaga. First described in 1935 from the type specimen collected near Marcapata east of Cusco.

Creamy-crested Spinetail *Cranioleuca albicapilla*
Peruvian Endemic. 17 cm. 2500-3800 m. An unmistakable spinetail with a prominent creamy-white crown. Found singly and in pairs, this species generally avoids moist montane forest and is found almost entirely in semi-arid woodland, being quite tolerant of disturbed cultivated areas. Feeds by hopping and clinging to the underside of branches in thick bushes, looking for insect. Often raises crest especially when agitated. The song is a loud, laughing, descending series of notes – *'kjeep kjeep kjeep kjeep kjeep,'* etc., lasting for 5-7 seconds. Most easily seen along the Inca Trail which follows the Cusichaca River near Llactapata ruins also on the Ollantaytambo side of Abra Malaga.

TYRANT FLYCATCHERS   Tyrannidae

The New World flycatchers are the largest New World family and pose notoriously tricky identification challenges. Many are simply difficult (ground-tyrants) and others, such as the genus *Tolmomyias*, identification depends on knowing their calls. They are found in all climates and habitats, from arctic snow line conditions to humid tropical rainforests, and are as equally diverse in behavior – from terrestrial to strictly arboreal. Most feed by fly-catching techniques which vary in style, from true aerial hawking, to perch-gleaning and hover-gleaning. Many that nest in temperate zones are migratory. They are mostly monogamous with sexes being similar in the majority of cases, although there are some exceptions. Songs are simple and usually given at dawn. Nest sites and constructions vary, from open cups on the ground to oval hanging nests with a side entrance in the canopy. Some nest in cavities and in some species both sexes incubate.

Sclater's Tyrannulet *Phyllomyias sclateri*
12 cm. < 2200 m. Shows yellow base to lower mandible. Found at the edges and in the canopy of sub-montane forest, often in clearings and lighter woodland dominated by alders (*Alnus* sp.). Seen singly or in pairs, sometimes with mixed feeding flocks. Hover gleans actively in foliage in the upper or outer parts of the tree canopy. Shivers and briefly lifts its wings. The distinctive call is a series of harsh emphatic notes. This species is at the northernmost limit of its range at Machu Picchu, but can be seen quite easily along the lower part of the road climbing to Machu Picchu Ruins. Named for Philip Lutley Sclater (1829-1913) English lawyer, ornithologist, founding editor of *The Ibis* and Secretary of the Zoological Society.

Black-capped Tyrannulet *Phyllomyias nigrocapillus*
11 cm. < 3100 m. Horizontal posture. Active. Easily identified by blackish cap, very yellow below, lack of tawny rump. Found in forest and borders in canopy or mid-story. Song a high pitched note followed by an even-pitched trill. Uncommon and best looked for at Machu Picchu along the railroad track.

Ashy-headed Tyrannulet *Phyllomyias cinereiceps*
10 cm. < 2450 m. Another confusing *Phyllomyias*. A large black crescent on the ear coverts. Occurs in the canopy, sub-canopy and borders of montane and sub-montane forest. Perches upright. Forages alone or in pairs with mixed species flocks catching flying insects among the outer branches of middle canopies. Call is a high pitched *'pseee'* given while foraging, followed by a high-pitched descending one-second trill. Can be seen along the railway track between Puente Ruinas and the Mandor Valley and on the grounds of the Machu Picchu Pueblo Hotel.

Marcapata Spinetail

Creamy-crested Spinetail

Sclater's Tyrannulet

Black-capped Tyrannulet

Ashy-headed Tyrannulet

Tawny-rumped Tyrannulet   *Phyllomyias uropygialis*
Length 10-11.5 cm. < 3600 m. Dark cap and conspicuous tawny rump. Found at the edges of humid montane and pre-montane forest, bushy clearings and isolated semi-humid patches of forest, sometimes in quite dry habitat. Found singly and in pairs, often accompanying mixed feeding flocks. Feeds by a combination of perch gleaning and hover gleaning, searching for insects on the surface of leaves and twigs. The call is a scratchy *'tzzit.'* The song consists of two wheezy notes, the first higher-pitched. Can be seen in forest patches along first part of the Inca Trail and scrubby semi-humid areas in the Sacred Valley of the Incas.

Plumbeous-crowned Tyrannulet   *Phyllomyias plumbeiceps*
11.5 cm. < 2200 m. Very similar to the preceding species but more yellow below. Note a black stubby bill. Found in the canopy and sub-canopy of pre-montane forest and forest borders. Usually with mixed flocks. Sallies and hover gleans twigs and leaves. Will occasionally lift a wing above its back but not constantly, like the *Leptopogon* flycatchers. Song is a loud sharp accelerating *'pip-pip-pip-pip-PIP-PIP PIPtrr.'* Can be seen along the railway track between Puente Ruinas and the Mandor Valley and on the grounds of the Machu Picchu Pueblo Hotel.

Yellow-bellied Elaenia   *Elaenia flavogaster*
16-16.5 cm. < 2000 m. Not particularly yellow bellied, despite its name. Usually shows an erect bushy crest parted in the center with a median white coronal patch. Found in lighter woodland, clearings and gardens. Quite active and conspicuous, found singly or in pairs, often at fruiting trees. Very vocal, giving a buzzy rasping *'breeer'* or *'reeek-kreep'* often repeated. At the upper elevational limit of its range at Machu Picchu. Rare, but may be expanding due to deforestation below Machu Plcchu.

White-crested Elaenia   *Elaenia albiceps*
13-14.5 cm. < 3300 m. The resident *urubambae* race is present in our area. Found alone or in pairs, active but shy. Inhabits edges of clearings, riparian thickets, gardens and orchards, and particularly likes pepper-trees *(Schinus molle)* and willows *(Salix chilensis)*. One of the common garden birds in the Sacred Valley of the Incas. The call is a buzzy *'bwreeee.'*

Small-billed Elaenia   *Elaenia parvirostris*
13-14 cm. < 3000 m. An austral migrant, present between April and September. Very similar to the preceding species and probably not safely told apart in the field. Shows a stubbier bill and a purer gray throat and breast. Usually shows a smaller third wing bar, but these differences are subtle. When on wintering grounds, usually encountered in clearings, bushy slopes and gardens but has been recorded in the canopy of humid forest where it undoubtedly goes unnoticed. The song is similar to that of the white-crested Elaenia – *'chi-brr'* and *'cheuu.'* May be encountered at all elevations, but rare, often in loose feeding aggregations.

Highland Elaenia   *Elaenia obscura*
17 cm. > 2700m. A large, dark-looking, elaenia with a short bill and rounded head. Looks pot-bellied and long-tailed. Inhabits undergrowth and clearings in humid montane and pre-montane forest and secondary woodland. Tends to remain more under cover than other elaenias, perching on the outer branches of trees in mid canopy and actively fly-catching. Feeds on berries also. Has a variety of calls – the most often heard being a rapidly delivered *'bureeeep.'*

Sierran Elaenia   *Elaenia pallatangae*
14-14.5 cm. < 3500 m. Conspicuous yellowish eye-ring with two yellowish-white wing bars. Underparts lemon yellow, with upper-breast and throat washed with olive gray. Inhabits borders of forest, shrubby clearings and secondary woodland. Seems to like alders in particular. Usually found alone, although several birds may congregate at a fruiting tree. Fly-catches and hover gleans from open perches for insects and small fruits. The call is a clear *'wheeoo.'* Common.

White-tailed Tyrannulet   *Mecocerculus poecilocercus*
11 cm. < 2500 m. Rump and upper-tail coverts conspicuously pale yellow, visible in flight. Inhabits canopy and borders of humid montane and sub-montane forest. Alone or in pairs with mixed feeding flocks, hopping along thin branches perch-gleaning. Often fans tail showing white edges and also spreads wings showing yellow rump. The call is a descending *'zee-zee-zee-zee,'* with the last two notes softer than the first two. At the southern edge of its range at Machu Picchu. Can be seen along the Urubamba River between Aguas Calientes and the Mandor Valley.

White-banded Tyrannulet   *Mecocerculus stictopterus*
11.5-12.5 cm. 2400-3600 m. Note conspicuous double white wing bars and superciliary. A pretty flycatcher found in and at the borders of humid montane and pre-montane forest, showing a preference for stands of alders. Found in pairs or small family groups, often associated with mixed feeding flocks. Perches horizontally much like a warbler, actively hover-gleaning and fly-catching among the outer branches of trees and bushes at low to mid-levels. The frequently heard call is a snarling, upward-slurred *'ssqueeyah'* repeated 3-5 times. Can be seen along the Inca Trail and at Abra Malaga.

White-throated Tyrannulet   *Mecocerculus leucophrys*
13-14 cm. 2500-4600 m. Long-tailed, with an upright posture. Distinctive white throat, throat feathers usually puffed out. Found in humid montane and elfin forest, as well as low bushes at tree line, and sometimes in isolated *Polylepis* groves. Found in pairs and small groups, invariably but not always associated with mixed feeding flocks. Often perches at middle heights and in the open. Feeds by hover-gleaning the underside of leaves and fly-catching, seldom returning to the same perch. The contact call is a sharp *'pit-pit-pit.'* The seldom-heard dawn song is a jumble of warbled phrases. A common bird at and just below the tree line along the Inca Trail and at Abra Malaga.

Ash-breasted Tit-Tyrant   *Anairetes alpinus*
13-13.5 cm. 3700-4600 m. Virtual Peruvian endemic (found also in La Paz, Bolivia). Globally threatened. Long, parted crest revealing a white patch on the hind-crown. Tail black, with conspicuous white outer tail feathers. Found exclusively in isolated patches of *Polylepis*, often mixed with *Gynoxys* bushes. Feeds on the outside of the canopy of very dense *Polylepis* trees and bushes, flitting and climbing on the outer branches. Found alone or in pairs. Both perches and hover gleans. Can be seen at Abra Malaga and other *Polylepis* groves in the Cordillera Vilcanota.

Yellow-billed Tit-Tyrant   *Anairetes flavirostris*
10-11.5 cm. < 2600 m. Dark iris and pale yellow lower mandible. Found singly, in pairs or small family groups, often with mixed flocks, generally in arid or semi-arid regions among scrub, bushes and riparian thickets. Feeds at the top of bushes and other low vegetation. The song is a series of rising and descending notes *'zeet-whee-whee-whee whee-wheee.'* Quite common in drier habitats.

Tufted Tit-Tyrant   *Anairetes parulus*
11-12 cm. > 3700 m. Iris pale yellow and bill all black. Occurs in humid forest borders and clearings but also in semi-arid isolated forest patches. In general prefers more humid habitat than the Yellow-billed Tit-Tyrant. In pairs or family groups often with mixed species flocks. Both perches and hover gleans, and makes short aerial sallies for insects. Flicks tail. Fairly common.

Unstreaked Tit-Tyrant   *Uromyias agraphia*
13 cm. 2700-3400 m. Peruvian endemic. The nominate *agraphia* subspecies is present. Elongated black crest. Found in the lower growth and borders of humid montane and elfin forest, especially where there are dense stands of *chusquea* bamboo. Found alone, in pairs or small groups, often with mixed flocks of warblers and hemispingus. Active, and makes upward sallies for insects and hover gleans the underside of bamboo leaves and twigs before dropping back into cover. Wags its tail from side to side. Calls include *'tzeeree,' 'tsit-trrrt'* and *'tsueweettsie.'* Uncommon at Abra Malaga.

Torrent Tyrannulet *Serpophaga cinerea*
11.5 cm. < 3300 m. Unmistakable. Concealed white coronal patch. Quite common along fast-flowing streams and rocky rivers in both humid and arid zones. Perches on stones and boulders out in streams and rivers or perched on overhanging branches, usually in pairs. It flicks its tail constantly, occasionally cocking it. Sallies upwards for flying insects but also runs across rocks etc. picking insects from the surface. Gives a sharp, loud *'chip'* note, often repeated several times. Common along the Urubamba River and its tributaries.

Subtropical Doradito *Pseudocolopteryx acutipennis*
11.5 cm. 3000 -3300 m. Small, slender and warbler like with buffy wing bars. Black bill (juveniles with pale lower mandible). Inhabits reed beds, corn and grain fields in agricultural areas. Breeds in the Cusco valley (sporadic?). Also an austral migrant May to September. Can be seen at Huacarpay Lakes. Migrant birds may turn up in riparian thickets. Call a dry *'tzik.'* Little known in the Andes and may be overlooked. Doraditos so named for their golden underparts; doradito, diminutive of Spanish dorado.

Rufous-headed Pygmy-Tyrant *Pseudotriccus ruficeps*
11 cm. 2000-3500 m. Distinctive and unlikely to be confused, but quite difficult to see. Found in dense, mossy undergrowth and on the borders of humid montane and pre-montane forest, usually alone or in pairs. Very active, making short flights between branches perch gleaning for insects on the underside of leaves. Wings whirr when the bird flies and it often snaps its beak. The song is a loud buzzy trill *'tzzzzzzzzzzzzzzzzzzeeeer'* lasting two seconds. Can be seen at Abra Malaga.

Bolivian Tyrannulet *Zimmerius bolivianus*
12 cm. < 2600 m. The brighter *viridissimus* subspecies is present at Machu Picchu. Note pale yellow edges to flight feathers and lack of wing bars. Found singly, rarely accompanying mixed species flocks, in the canopy and borders of humid montane forest, mature secondary woodland and adjacent wooded clearings. Feeds mostly on mistletoe berries. Can be seen on the lower slopes of Machu Picchu. Genus named after John Todd Zimmer (1889-1957) US ornithologist who wrote extensively on Peruvian birds.

Marble-faced Bristle-Tyrant *Phylloscartes ophthalmicus*
11.5 cm. < 2000 m. The *ottonis* race is present at Machu Picchu. Preferred habitat is the lower and middle growth of humid sub-montane forest and borders. Usually found singly or in pairs accompanying mixed species flocks. Perches vertically on exposed branches, frequently flicking a wing over its back, and sallying out to pick insects off leaf surfaces. The song is a high-pitched *'psee-ee-e-e-tititi.'* Likely only in the lower parts of the Machu Picchu Sanctuary.

Mottle-cheeked Tyrannulet *Phylloscartes ventralis*
12 cm. < 2400 m. Slim, long tailed and active. Cocks tail and droops wings. Found in the canopy and borders of humid sub-montane forest and well developed secondary woodland. Singly or in pairs with or away from mixed species flocks. Perch gleans on outermost twigs and leaves in the open. The song is a quick *'whik-whik-whi-i-i-i-i-r whik-whik.'* Frequently encountered along the railway track, between Puente Ruinas railway station and Mandor.

Streak-necked Flycatcher *Mionectes striaticollis*
13 cm. < 3000 m. Upright posture. Head and neck are gray, with a prominent white spot behind the eye. Does not show wing bars. Rather common in the lower growth and sub-canopy of montane and pre-montane forest and secondary woodland. Often found with other species at fruiting trees, and sometimes joins mixed tanager flocks, but just as likely to be encountered alone. Sits quietly and forages out, hovering to pick off small fruits. Quite common at lower elevations within the Machu Picchu Sanctuary.

Many-colored Rush-Tyrant *Tachuris rubrigastra*
10 cm. 3100-4600 m. A colorful, unmistakable flycatcher. Inhabits thick reed beds (*Scirpus* and *Typha*). Usually encountered in pairs or small family groups. Restless and agile, frequently flicking its wings and tail. Perch gleans low in the reeds, often at water level, on exposed mud or pond weed. The call is a nasal *'piuh-piuh'* repeated up to five times. Common around edges of lakes in the Cusco area, Huacarpay and Huaypo Lakes for example.

Slaty-capped Flycatcher   *Leptopogon superciliaris*
13.5 cm. < 2000 m. Upright posture. Found in the lower and middle growth of humid sub-montane forest, almost always with mixed species flocks. A perch on exposed branches and hover gleans insects off the underside of leaves. Flicks one wing over its back, like others in this genus. The song is an often heard, well defined *'streeer di-i-i-i-i-rr.'*

Inca Flycatcher   *Leptopogon taczanowskii*
12-13 cm. < 2800 m. Peruvian endemic. Upright posture and tawny wash to breast. Fairly common in middle and lower growth of montane and pre-montane forest and forest edge. Follows mixed feeding flocks but also encountered singly. Characteristically, it lifts its wings over its back repeatedly. Makes short sallies for flying insects or hover-gleans from the underside of leaves. Will also eats small fruits. The call is an often-repeated, explosive *'tzeet,'* repeated every second or so. Best looked for along the railway track between Puente Ruinas and the Mandor Valley.

Scale-crested Pygmy-Tyrant   *Lophotriccus pileatus*
10 cm < 2100 m. Long crest broadly edged with rufous. Understory of humid montane forest. Inconspicuous and quiet and often first detected by its often given call – ringing descending buzzy trills *'dzzeer- dzeer-dzeer'* or just *'turreee.'* Uncommon in our area but can be found near Machu Picchu ruins.

Black-throated Tody-Tyrant   *Hemitriccus granadensis*
9-10 cm. 2000-3000 m. The *pyrrhops* race is present at Machu Picchu. Buff lores and ocular area. Usually found in lower growth of montane forest in small openings and clearings, alone or in pairs, seldom joining mixed species flocks. Moves in short hops and often stays in the same area for a long time. Perches in the open and sallies upwards to pick insects off the underside of leaves. The song is a fast, fairly loud *'whiiddik'* and a sharp *'whiip-whiip-whiip.'* Can be seen on the Inca Trail and at Abra Malaga.

Common Tody-Flycatcher   *Todirostrum cinereum*
9-9.5 cm. < 2000 m. Long flat bill. Flight feathers are black edged bright yellow. Found in shrubbery, lighter woodland, forest borders and gardens. Avoids heavily forested habitat. Very active, usually in pairs, hopping and fluttering through branches with its tail perpetually cocked. Hover gleans on the undersides of leaves and sometimes fly-catches. The song is a sharp *'tweet'* or a trilled *'treeeeet,'* repeated several times. Easily seen around the Puente Ruinas railway station and on the grounds of the Machu Picchu Pueblo Hotel.

Yellow-olive Tolmomyias (Flycatcher)   *Tolmomyias sulphurescens*
14 cm. < 2000 m. A large headed, flat-billed flycatcher. The genus *Tolmomyias* is a difficult group of flycatchers to identify, but possible confusion species are found only in the Amazon lowlands. The race at Machu Picchu is the *peruvianus* foothill race. The throat is flecked grayish-white, breast and flanks olive, becoming pale yellow on the belly. The wings are dark with two prominent yellow wing bars and edgings to the flight feathers. Inhabits the sub-canopy and borders of humid pre-montane forest and open areas with some tall trees. Forages at medium heights seldom cocking its tail. Found alone or in pairs, also frequently following mixed species flocks, where it is quite deliberate and slow in its movements. Jumps at foliage looking for insects, which it often chases. Sometimes eats small berries. The call is a series of rising sharp *'tszreet'* notes. Can be seen around the site museum at Puente Ruinas.

Bran-colored Flycatcher   *Myiophobus fasciatus*
12-12.5 cm. < 2000 m. In our area above reddish brown, crown reddish with hard to see semi-concealed orange-yellow coronal patch. Wings dusky with two prominent buff wing bars. Usually found alone or in pairs in forest edge, secondary growth or shrubby clearings. Often inconspicuous, despite its preference for open habitats, perching low and usually not in the open. Sallies to foliage as well as fly-catching for insects but eats some small fruits. The call is a whistled *'jleeb'* or *'jluub.'* Dawn song is a repetitious *'djlee-dluu djee-dlu.'* Rare at Machu Picchu, where it is an austral migrant from the south likely between May and September. Has been seen along the railway track between Puente Ruinas and the Mandor Valley.

Cinnamon Flycatcher   *Pyrrhomyias cinnamomea*
11-13 cm. 1800-3400 m. One of the most familiar flycatchers of the forest. Perches in an upright manner. Wings and tail black, wings showing two conspicuous rufous wing bars and wing panel. Found in clearings and borders in humid montane and pre-montane forest and mature second growth, often at the sides of tracks and trails and thus easily seen and familiar. Perches on exposed horizontal branches at mid-levels, usually in pairs and not with mixed species flocks. Sallies out to hawk for insects, returning repeatedly to the same exposed perch. Tame. The often-heard call is a buzzy *'trrrrrt'* given throughout the day.

Cliff Flycatcher   *Hirundinea ferruginea*
16-18 cm. < 2200 m The *sclateri* race is present at Machu Picchu. A large slender flycatcher that perches horizontally. Tail dark with cinnamon rufous at base and rufous rump, conspicuous in flight. Large amounts of rufous visible in the wing in flight. Found close to cliffs and in rocky canyons, exposed granite outcrops, road and railway cuttings. Usually in pairs perched on rock ledges or small branches on rock faces. Sallies out for quite long distances to hawk for insects in a swooping, swallow-like fashion, returning to the same perch. Quite vocal and makes a variety of high-pitched calls. Can be seen on the vertical granite cliff faces near Machu Picchu.

Ochraceous-breasted Flycatcher   *Nephelomyias ochraceiventris*
12-13.5 cm. 2200-3700 m. Yellow-orange median crown patch. Breast and throat ochraceous (earthy yellow = ochre). Wings show two distinctive buffy or whitish wing bars. Inhabits humid montane and elfin forest. Can be seen singly but also up to four or five birds together. Often encountered in mixed feeding flocks of tanagers etc. Feeds by fly-catching from the tips of branches in the middle and upper strata of trees. Sometimes perch gleans. Quite conspicuous and easy to observe. Can be seen along the Inca Trail between Sayacmarca and Phuyupatamarca and at Abra Malaga.

Alder Flycatcher   *Empidonax alnorum*
14 cm < 3600 m. Boreal migrant present September to April. Note whitish eye-ring, whitish wing bars. Often flicks tail upwards. Call a sharp *'pip.'* Found in forest edge, riparian thickets, and agricultural areas. During migration can turn up anywhere there is light vegetation.

Olive-sided Flycatcher   *Contopus cooperi*
18 cm. < 2300 m. A boreal migrant from North America, present in Peru from September to May. Two indistinct pale grayish wing bars. A diagnostic tuft of white may protrude from behind the wing onto the sides of the rump. Occurs at borders of montane forest and in clearings with scattered trees. Perches high on exposed (often dead) branches. Sallies out, often for long distances, for flying insects and returns habitually to the same perch. The frequently heard call is a loud *'pip-pip-pip.'* Uncommon at Machu Picchu.

Smoke-colored Peewee   *Contopus fumigatus*
17 cm. < 2800 m. The *ardosiacus* race is present at Machu Picchu. A large, upright, uniform slate-gray flycatcher, with a prominent bushy crest. Found in borders and clearings of montane forest with scattered trees. Alone or in pairs, perches in the open, often changing sites. Sallies out quite long distances for aerial insects, returning sometimes to the same perch, shivering its tail on alighting. Tame. The frequently heard call is a loud *'pip-pip-pip.'* The less frequently heard song is a hoarse *'zur-zur-zur-zur-zur.'* Fairly easily seen along the railway track between Aguas Calientes and the Mandor Valley.

Western Wood Pewee   *Contopus sorididulus*
14.5 cm < 2200 m. Rare boreal migrant September-April. Winter range poorly known, can be confused with similar Eastern Wood-Pewee. Note narrow wing bars and lack of eye-ring. Call *'bureeer.'* Can be seen along the railroad track and in the Mandor Valley at the right time of year.

Olive Flycatcher   *Mitrephanes olivaceus*
12 cm. < 2000 m. Conspicuous crest. Above mostly olive, underparts buffy-olive with a slight yellowish wash on the belly. Inhabits undergrowth and middle levels of humid pre-montane forest. Perches upright, more or less in the open, and sallies out short distances to catch flying insects, often returning to the same perch with a characteristic shivering of the tail on alighting. Often encountered with mixed species flocks. This species is at the upper elevational limit of its range and only likely to be encountered in the lower Mandor and Aobamba Valleys.

Black Phoebe   *Sayornis nigricans*
17.5 cm. < 2800 m. The *latirostris* race is present at Machu Picchu. Mostly black. Central belly, two wing bars, large wing panel and outer-edge of outer tail feathers white. Commonly found along rocky streams and rivers. Confiding and easily observed. Usually in pairs, perched on boulders and rocks in swift moving streams and rivers, sometimes on bridges. Jerks tail upwards. Feeds by sallying into the air after flying insects. Quite common along the lower Urubamba River and its tributaries.

Vermillion Flycatcher   *Pyrocephalus rubinus*
14-15 cm. < 2200 m. A rare austral migrant present between April and October. Male unmistakable. Cap, and underparts vermillion, upperparts, tail and mask through the eye sooty black. At Machu Picchu solitary and found in clearings, open areas and riparian thickets usually conspicuously perched in the open on the lower branches of trees, fences, walls etc. Often flicks its wings and spreads and pumps its tail. Sallies, and often drops to the ground, from a perch for insects, quickly fluttering back to the same perch. Often aggressive toward other birds. The call is a sharp *'peent.'*

Andean Negrito   *Lessonia oreas*
12.5 cm. 3100-4600 m. Male black with a rufous back. Female blackish brown, chin white with the back and sides of breast a dull, dusky rufous. Inner webs of flight feathers in both sexes whitish, visible in flight. Inhabits wet open areas with short grass or even very barren ground and the muddy edges of lakes. Always associated with water. Terrestrial and found in pairs or small family groups. Often perches on tussocks and sallies out close to the ground to catch flying insects. Also, dashes across the ground in search of prey. Common at lake edges in the Cusco area.

Rufous-tailed Tyrant   *Knipolegus poecilurus*
14.5 cm. < 2300. Sexes similar. Iris red. Tail dark with inner webs rufous – conspicuous in flight. Upper tail coverts rufous. Found low in vegetation at the borders of humid montane forest and in forest clearings with some tall herbs and small bushes. Does not associate with mixed species flocks. Perches in the open on exposed branches often raising, then slowly lowering its tail. Drops to the ground or makes short sallies in search of prey. Uncommon.

Plumbeous Black-Tyrant   *Knipolegus cabanisi*
16 cm. 1800-2700 m. Male: iris red and bill blue-gray. Plumage is uniform plumbeous gray. Wings and tail blacker. Female: iris red-brown and bill dark. Tail dusky with edges of feathers rufous-cinnamon. Inhabits lower to middle strata of humid montane and sub-montane forest, often at the edge of clearings or trails. Inconspicuous and quiet, usually alone, perched on exposed branches and twigs in the understory. Perches upright with a slightly hunched posture and constantly shivers the tail. Has a swooping display flight. Named for German ornithologist Jean Louis Cabanis (1816-1906) founding editor of the *Journal für Ornithologie*.

White-winged Black-Tyrant   *Knipolegus aterrimus*
16-17 cm. < 3000 m. Both sexes show a dark iris and dark bill. Male has a white band the across base of the primaries, not visible at rest but conspicuous in flight. Female rump rufous, tail dark above, and rufous below with a dark tip. Inhabits lighter woodland and scrubby slopes but also found in clearings in more humid forest. Alone or in pairs, perches with an upright stance in the open at the top of bushes and small trees. Sallies from a perch to fly-catch or onto vegetation or the ground in search of prey. The male has a swooping display flight. The seldom-heard song is a thin *'chit-zer.'* A conspicuous bird quite common at the Machu Picchu ruins.

Spectacled Tyrant   *Hymenops perspicillaus*
16 cm 3350 m. Rare austral migrant. Distinctive. One record from Lake Piuray. Found in marshes and wet grassland. Perches conspicuously on fencepost or bushes.

Little Ground-Tyrant   *Muscisaxicola fluviatilis*
13.5 cm. < 3100 m. A bird of the Amazon but does reach the Andean foothills. Sometimes higher – austral migrant? (Huacarpay Lakes). Inhabits river margins or manmade clearings in lowlands, open areas in the Andes. Perches on weeds or fallen branches and rocks. Like all ground-tyrants has an upright stance and white edges to tail feathers. Note small size and buffy breast contrasting with white belly. Rare.

Spot-billed Ground-Tyrant   *Muscisaxicola maculirostris*
15 cm. 2000-4000 m. Resident all year round but some elevational movements. Upper-parts sandy brown with dark line through the eye and a whitish supercilium. "Spot" on bill formed by yellowish base of lower mandible not usually visible in the field. One of the easier ground-tyrants to identify in this otherwise difficult genus, being smaller and browner with a shorter tail than its congeners. Inhabits open bushy country and rocky slopes always close to ravines or rock walls, often among stone walls and fields. Found alone or in pairs, sometimes several together in the austral winter. Perches on bushes, banks, walls or on the ground with an upright stance typical of the genus and often quivers its tail. Can be seen at Huacarpay Lakes.

Taczanowski's Ground-Tyrant   *Muscisaxicola griseus*
17-18 cm. 3300-4800 m. No crown patch. Resident year round. Broad white supercilium extending past the eye, with the area below the eye whitish. Found on level or gently-sloping, grassy soil with some rocks. Habits and display similar to other ground-tyrants, though perhaps with a less upright stance. In the non-breeding season (April-September) often in loose flocks, commonly with other species of ground-tyrants. Common at Abra Malaga and the Cordillera Vilcanota. Very similar to Cinereous Ground-Tyrant.

Puna Ground-Tyrant   *Muscisaxicola juninensis*
16 cm. 3800-4800 m. Resident year round. A confusing species. Strongly tinged brown on rear of crown (but not a coronal patch). Short white supercilium. Inhabits high puna grassland close to rock outcrops, cliffs, and around lakes, marshes and bogs. Aerial display and habits as in other ground-tyrants. Only found at higher elevations. Fairly common at Abra Malaga and the Cordillera Vilcanota.

Cinereous Ground-Tyrant   *Muscisaxicola cinerea*
15.5 cm. 4000-4500 m. An austral migrant present between April and September. Very like Taczanowski's Ground-Tyrant but pale bluer-gray above and white below. It also differs in its smaller size and short supercilium which extends to the eye only. Prefers short matted vegetation and rocky outcrops on hillsides. Habits and display as for other members of the genus. Often seen in loose flocks. Uncommon at high altitudes. Cinereous = ash colored.

White-fronted Ground-Tyrant   *Muscisaxicola albifrons*
24 cm. 3800-4800 m. Resident year round. The largest ground-tyrant, and therefore fairly easy to identify. Very upright stance. Long wings reach the tail tip. Note large size, plain crown and conspicuous pale edges to wing coverts. Always on bogs with matted grass and cushion plants, sometimes wandering onto drier slopes, where it often hops onto tussocks or rocks to survey the terrain. Habits and display as in other members of the genus, forming loose groups in the non-breeding season. Uncommon.

Ochre-naped Ground-Tyrant   *Muscisaxicola flavinucha*
20 cm. 3800-4700 m. An austral migrant present between April and September. Almost as big as the preceding species and with very long wings reaching to tail tip A conspicuous patch on lower hindcrown yellow-ochre. Broad frontal area and short supercilium white. When in Peru, prefers level bogs and lake-sides with short, matted vegetation. Habits and display as for other members of the genus. Uncommon.

Rufous-naped Ground-Tyrant   *Muscisaxicola rufivertex*
16-17 cm. 3000-4500 m. Resident year round. Note the very upright posture, long bill with decurved tip. Above pale gray contrasting with blackish rump, with a conspicuous chestnut patch on the hindcrown (not nape). Inhabits grasslands and rocky hill slopes, plowed fields and ravines. Found singly, in pairs or small groups in the austral winter at lower altitudes. Feeds on quite flat areas, often near water, where it surveys the terrain from the ground or a slightly elevated perch and dashes after insects. Aerial display is as in the preceding species. Fairly common at Huacarpay Lakes in winter.

White-browed Ground-Tyrant   *Muscisaxicola albilora*
17 cm. 3000-4200. An austral migrant present between April and September. Another confusing member of the genus. Upper-parts gray brown, forehead brown with a broad diffuse (not crisp) tawny hindcrown. Supercilium longer and brighter than others in the genus with black eye line. Occurs on open puna grassland and pastures, around lakes and ponds. On migration often seen in loose flocks and highly mobile. Habits and behavior as for other ground-tyrants. Fairly common in winter.

Black-billed Shrike-Tyrant   *Agriornis montana*
23 cm. 3000-4500. Bill slender and black (lower mandible yellow in juvenile). Central tail feathers dusky, the rest bright white and very noticeable in flight. Shows a pale iris. Inhabits open country, dry bushy hillsides, open puna grassland with boulders, stone walls and ruined buildings etc. Solitary. Perches in the open on rocks, trees, bushes and rock faces. Feeds on the ground. Searches for insects by gliding from rock to rock, twisting abruptly to pounce on prey. The call is a clear, whistled *'wheeeuu.'* The most common and widespread shrike-tyrant.

White-tailed Shrike-Tyrant   *Agriornis albicauda*
25 cm 3000-4300 m. Rare. A few scattered records. Hard to tell from the previous species without great care. Note stout bill with larger hooked tip with pale lower mandible (but beware immature of the preceding species), large size and much more, well defined black streaks on throat. Browner below. Large head. Not any more "white-tailed" than the previous species. Habitat similar to previous species – may like drier habitats.

Gray-bellied Shrike-Tyrant   *Agriornis micropterus*
24 cm. 3000-4100 m. Status unclear, but rare and possibly declining austral migrant. Differs from the previous two species by largely dark tail and buffier tones to underparts. Well defined streaked throat and pale lower mandible. Has been seen near Huacarpay Lakes.

Streak-throated Bush-Tyrant   *Myiotheretes striaticollis*
23 cm. 2000-3700 m. Wings dusky with rufous edgings. Tail cinnamon, with a broad black terminal band. In flight shows much rufous in the wings and tail. A bird of open country, found along hedgerows, near small farms, open wooded ravines, etc. When found in humid montane forest, always associated with large clearings, landslides and cuttings. Tame. Perches conspicuously on large branches, dead trees and tall bushes, from where it sallies forth long distances for flying insects, often returning to the same perch. Jumps to the ground for prey also. The call is a clear, whistled *'pseeeeeee.'* Can be seen around the Machu Picchu ruins.

Smoky Bush-Tyrant   *Myiotheretes fumigatus*
20 cm. 2300- 4200 m. Rare. A dark, nondescript but vocal bush-tyrant at the southern edge of its range here. In flight shows a large amount of rufous in the wing with a dark terminal band on the secondaries. Inhabits humid montane and sub-montane forest, often near clearings. Found singly or in pairs, sometimes with mixed feeding flocks, where its behavior can be very non-flycatcher like and confusing, hopping along branches and banging grubs on branches in the manner of a thrush. When not with flocks, its behavior is more typical, perching on open branches inside bushes and trees at mid to upper levels from where it sallies forth for flying insects or hover-gleans mossy branches and epiphytes. Dawn song is a monotonous series of whistles *'pew-pew-pew-pew'* three notes per second. Can be seen near San Luis at Abra Malaga.

Rufous-bellied Bush-Tyrant   *Myiotheretes fuscorufus*
19 cm. 2000-3400 m. Wings blackish, with two broad rufous wing bars and edgings to inner flight feathers. Shows a lot of rufous in the wing in flight. Found at the borders of humid montane and pre-montane forest and the edge of clearings, sometimes in alders. Encountered alone, in pairs or small family groups in the mid and upper strata of trees where it catches flying insects by hawking or sally gleans in foliage. Uncommon and seldom encountered at Machu Picchu or Abra Malaga.

Red-rumped Bush-Tyrant   *Cnemarchus erythropygius*
23 cm. 3000 4600 m. A large, unmistakable bush-tyrant with a white fore-crown merging into a hoary gray crown. Above, brownish slate with a contrasting red rump. Found at the tree line in open habitat with scattered trees and shrubs, also in *Polylepis* groves and occasionally into surrounding puna grasslands. Found singly or in pairs perched conspicuously on the tops of bushes, small trees, fence-posts and boulders. Takes most of its prey by hopping to the ground, but does occasionally fly-catch. Generally silent but gives a scratchy whistle. Generally scarce. Can be seen at Abra Malaga.

Rufous-webbed Bush-Tyrant   *Polioxolmis rufipennis*
21 cm. 3000-4600 m. Above and below smoky gray. Belly and vent whitish. Iris pale. In flight shows extensive rufous in the wings and tail with a broad dark terminal band on the secondaries. Found in open bushy country, puna grassland with occasional shrubs and boulders, farms and gardens and in *Polylepis* groves. Alone or in pairs, perched at the top of a boulder or tree from where it drops to the ground for prey, fly-catching less frequently. Often hangs and glides in aerial updrafts like an American Kestrel, sometimes several together. The call is a high pitched *'psweee.'* Uncommon.

Crowned Chat-Tyrant   *Ochthoeca frontalis*
12.5-13 cm. 2800-3600. In this area the *spodionota* subspecies. Garcia & Moreno *et al.* (1998) suggested that genetic and plumage differences warranted species recognition (Kalinowski's Chat-Tyrant). Frontal area golden-yellow, long white supercilium. Wings with two, sometimes only one, rufous wing bars. Found in understory and mid-levels of humid montane and elfin forest, shrubby clearings and edges. Usually encountered at or just below the tree line. Can be overlooked as it perches low in undergrowth, usually alone. Does not join mixed species flocks. Not very vocal but sometimes gives a thin trill *'trrt-trrrrr.'* Perch or hover gleans for insects in dark undergrowth. Can be encountered throughout the area where suitable habitat is present.

Golden-browed Chat-Tyrant   *Ochthoeca pulchella*
12-13 cm. < 3100 m. Here the nominate *pulchella* subspecies. Long yellow supercilium paler behind the eye. Two prominent rufous wing bars. Inhabits the lower strata of humid montane and sub-montane forest in dark undergrowth close to the ground, often where there is an understory of *chusquea* bamboo. Found alone or in pairs, sitting quietly and fluttering from perch to perch in search of invertebrates. Not a strong flier and extremely difficult to see. The call is a descending trill *'treeeeeee,'* or a more complex, longer version of this. Bill snaps frequently. Can be seen, with luck, near Intipunku on the Inca Trail not far above Machu Picchu.

Slaty-backed Chat-Tyrant   *Ochthoeca cinnamomeiventris thoracia*
13 cm. < 3300 m. Some consider the race *thoracia* found here a separate species – Maroon-belted Chat-Tyrant. (See Garcia & Moreno 1998). Dark, slate-gray with a distinctive white supercilium. Broad band of dark maroon chestnut across the breast. Found in the undergrowth and middle strata of humid montane and sub-montane forest, often at borders and always associated with running water and waterfalls, usually mountain streams with luxuriant growth. Seen singly or in pairs, not with mixed flocks, perched low at the edge or just inside the forest, making short sallies for aerial insects. The often-heard, high-pitched call is a long drawn out fruiteater like *'tseeeeeeyeeeee.'* Can be seen near the waterfall in the lower Mandor Valley at Machu Picchu.

Rufous-breasted Chat-Tyrant   *Ochthoeca rufipectoralis*
12.5-13 cm. 2300-4000 m. The *rufipectoralis* race is present at Machu Picchu. Chin and throat gray, lower throat and breast bright rufous, belly white. The wings are plain without a wing bar in this race. Found at the edge of humid montane and elfin forest, shrubby clearings and secondary growth. Sometimes in semi-humid isolated patches of woodlands in ravines. Perches conspicuously with an upright, slender posture, alone or in pairs, at middle levels and in the sub-canopy. Feeds by making short sallies for flying insects and also perch gleans the upper-sides of leaves. May join mixed species flocks briefly. The call is a soft *'chic-chica-chic.'* A familiar bird of the forested areas. Quite common along the Inca Trail, at Abra Malaga and in the semi-humid valleys flowing into the Sacred Valley of the Incas.

Brown-backed Chat-Tyrant   *Ochthoeca fumicolor*
15-16 cm. 2500-4200 m. The *berlepschi* race is present at Machu Picchu. Narrow dirty white supercilium. Shows two rufous wing bars, the lower one more prominent. Found in semi-open areas, borders of montane and elfin forest and also in isolated *Polylepis* groves. Solitary and conspicuous perched at the top or on thick branches of small trees and bushes. Also, perches readily on fence posts. Sallies from a perch into the air for flying insects. Will also jump to the ground to catch prey, returning to a different perch. Spreads wings and tail on landing, also flicks tail when alarmed. Call is a soft *'pseeuw.'* Can be seen near Llulluchapampa along the Inca trail and commonly at Abra Malaga.

d'Orbigny's Chat-Tyrant   *Ochthoeca oenanthoides*
15-16 cm. 3200-4200 m. Long broad supercilium is white. Throat flammulated whitish. Rest of underparts cinnamon rufous, paler on the vent and mid-belly. The white edges of the outer tail feathers are prominent in flight. Found in bushes and low trees in ravines at or above the tree line, often venturing out into rocky hillsides and the banks of streams with little or no vegetation. Also, frequents *Polylepis* groves where it is sometimes found in the same habitat as the Brown-backed Chat-Tyrant. Usually encountered alone or in pairs, perched upright on dead branches, bushes and fence posts from where it flies down to the ground to pick up insects, returning, more often than not, to the same perch. Can be seen near *Polylepis* at Abra Malaga. Named for Alcide Charles d'Orbigny (1802-1857), French naturalist and paleontologist who travelled widely in South America.

White-browed Chat-Tyrant   *Ochthoeca leucophrys*
14.5-15 cm. 2400-4200 m. The plain-winged *urubambae* race is present in this area. Broad frontal area white and long white supercilium. Outer tail feathers edged white. A bird of drier, more arid habitat than the preceding chat-tyrants. Often in quite xerophytic scrub, avoiding humid forest. Prefers ravines and scrubby slopes, often around small farms and hedgerows. Perches upright on the top of a bush from whence it sallies forth for insects, or drops to the ground for the same. Common around Cusco.

Piratic Flycatcher   *Legatus leucophaius*
15 cm. < 2000 m. Throat and breast whitish with obscure dusky brown streaking. Lower belly and vent pale lemon yellow. Inhabits open forest and clearings with tall trees. Solitary and usually perches high in the canopy on exposed snags and branches, often giving its distinct song, a high pitched down slurred *'tee-uu.'* The name of this bird derives from its habit of usurping the domed or pendant nests of a variety of birds including oropendolas and other flycatchers. The eggs of host species are always ejected from the appropriated nest. Seen in the Mandor Valley, but rarely.

Social Flycatcher   *Myiozetetes similis*
16.5-17 cm. < 2000 m. Olive above, tail dusky, wings same as the back, with some indistinct pale buff edging to the flight feathers. Inhabits a variety of habitats, including shrubby clearings, gardens, forest edge and lighter woodland canopy and borders. In pairs or small groups at all levels, from the ground to tree tops, on exposed perches. Often drops to the ground to catch prey. Noisy and active. Gives a variety of harsh calls, the commoner being *'kreee-you'* and *'kre-kree-kree-kree.'* At the limit of its altitudinal range of Machu Picchu, and only likely to be found below 2000 m. Can be seen along the railway track between Puente Ruinas and the Mandor Valley.

Great Kiskadee *Pitangus sulphuratus*
22 cm. < 2000 m. Vocal but rare in our area. Large. Lots of rufous in wings and tail. Large dark, heavy bill. Likes lighter woodland and forest edge. Forages low and perches conspicuously. Call *'keek-ker-deer.'* Can be seen at and near the Machu Picchu Pueblo Hotel.

Lemon-browed Flycatcher *Conopias cinchoneti*
16 cm. < 2000 m. Underparts entirely bright yellow, a little duller on the flanks. Found at edges and in clearings in humid pre-montane forest. Pairs or family groups perch conspicuously in the open on exposed branches or tops of leaves at high and mid-levels. They move about constantly and cover large distances when foraging, and are quite noisy, uttering a high-pitched *'whee-ee-ee-e whedidi-dididi.'* Uncommon.

Golden-crowned Flycatcher *Myiodynastes chrysocephalus*
20-21 cm. < 2800 m. Crown brownish-gray with a semi-concealed yellow coronal patch. Vague rufous edgings to feathers of wings and tail. Found in clearings, along rivers and at the edges of humid montane and sub-montane forest with scattered tall trees. Usually seen alone or in pairs perched in the open at middle heights. Quite vocal and noisy, the commonest call being a loud *'skeee-uuu,'* often repeated. Can be readily seen along the banks of the Urubamba River at Machu Picchu.

Streaked Flycatcher *Myiodynastes maculatus*
20-21 cm. < 2000 m. Large, blackish and heavily streaked *solitarius* race present, but also less boldly streaked and paler austral migrant to be expected. Rump and tail rufous-brown. Found in secondary woodland, forest edge and along rivers. Usually encountered alone and quite conspicuous. Perches at mid-heights for protracted periods, flying directly to another perch occasionally. Often noisy, giving *'chup'* or *'eechup'* notes. Has a surprisingly melodic and seldom-heard dawn song – *'whee-cheeder-eeee-whee.'* Rare at Machu Picchu.

Crowned Slaty Flycatcher *Empidonomus aurantioatrocristatus*
17-18 cm Austral migrant November to February in lowland once recorded at 3000 m at Huacarpay Lakes near Cusco. Black crown and semi-concealed golden crown patch. Could turn up anywhere as a migrant but prefers canopy and edge or lighter vegetated area. Usually silent in Peru.

Tropical Kingbird *Tyrannus melancholicus*
20-22 cm. < 2500 m. Head gray with dusky ear-coverts, bright yellow underparts. The ubiquitous flycatcher of the neotropics and a familiar bird in open areas of the lowlands. Colonizes man-made clearings rapidly and Machu Picchu is no exception. Found in clearings, lighter woodland, gardens and along large rivers, avoiding thickly forested areas. Perches in the open on exposed snags, branches and posts, as well as on phone wires etc. Makes long sallies for flying insects, often returning to the same perch. Usually solitary. The often-heard call is a high-pitched *'trrrrrrrrr,'* lasting about one second. Common at Machu Picchu.

Fork tailed Flyctacher *Tyranus savana*
38-40 cm (male), 28-30 cm (female). Austral migrant seen on passage February-April or August-October. Gray and white with distinctive long tail and contrasting dark cap. Likes open areas, fields and pastures, lake edge. Mainly insectivorous but will eat small fruits. Generally silent in Peru and rare on passage.

Eastern Kingbird *Tyrannus tyrannus*
19-20 cm. < 3300 m. Boreal migrant September to April. Often in flocks below 2000 m but lone strays higher. White tip to tail. In canopy of forest, forest edge and second growth. Call a high pitched *'tzeee.'* Uncommon.

Rufous Casiornis *Casiornis rufa*
17 cm. < 2000 m. Rare austral migrant to southern Peru and recorded once at Machu Picchu. Mostly rufous, slender bill with pink base. Likes river edge and second growth forest and gardens. Mostly silent. May be overlooked during migration.

Dusky-capped Flycatcher   *Myiarchus tuberculifer*
17.5 cm. < 3200 m. The *atriceps* race is in this area. Crest sooty-black half raised. Wings with only a narrow, faint bar on the greater coverts. Throat and breast pale gray, rest of underparts clear lemon-yellow. Found at the edge of humid montane and pre-montane forest, clearings, and secondary growth. Also, in semi-humid side valleys of the Sacred Valley. Perches at mid-levels conspicuously, usually alone, hawking for insects. Will sometimes accompany mixed species flocks. The often-delivered call is a wheezy *'freeeer.'*

Pale-edged Flycatcher   *Myiarchus cephalotes*
18 cm. < 2600 m. Bill all dark. Shows three well-marked, narrow wing bars and pale grayish white edges to the tertials. Tail feathers dusky, outer feathers edged paler. Found in humid pre-montane forest edge, sometimes up into humid montane forest, in clearings and secondary growth. Habits similar to the preceding species. The call is a loud *'peeuur,'* which is often repeated.

COTINGAS   Cotingidae

A large varied family closely related to the manakins and tyrant flycatchers. Many are colorful and have crests. Mostly frugivorous, cotingas may consume enough food for the day in less than an hour. They spend large amounts of time motionless in the canopy or sub-canopy of the forest. In some species several males congregate at display "leks" where they vocalize loudly. Nests vary from species to species and only the female incubates. Both parents feed the young.

Band-tailed Fruiteater   *Pipreola intermedia*
18.5-19 cm. 2000-3000 m. Male: Black head. Green tail with black sub-terminal band and white tip. Female: very different having a green head and breast like the back and tail. Found in lower and middle levels of humid montane and pre-montane forest and forest edge. Usually alone or in pairs, often at fruiting trees. The song is a very high pitched trill – *'tsi-tsi-tsi-tsiiiiiii'* lasting four or five seconds.

Barred Fruiteater   *Pipreola arcuata*
22-23 cm. 2000-3500. A large impressive fruiteater. The *viridicauda* race present in the Machu Picchu Sanctuary has a pale yellowish iris. Tail black and olive-green with broad black sub-terminal band and whitish tip. Wings with large pale yellow spots on the greater coverts and tertials. Inhabits humid montane forest and forest borders. Found singly and in pairs. Does not join mixed flocks, but often accompanies other species at fruiting trees. The song is a high pitched, thin *'see-ii-ii-ii-ii-ii-ii'* lasting for 3-4 seconds.

Masked Fruiteater   *Pipreola pulchra*
18 cm. < 2400 m. Peruvian endemic. Male: lower throat and center of breast orange. Rest of underparts clear yellow with green streaking on the sides and flanks. Female: bright green above with underparts green narrowly streaked with yellow. Both sexes have gray legs and show a yellowish-white iris. Inhabits humid pre-montane forest and mature second growth woodland. Lethargic, sits still for long periods. Like other fruiteaters, often found alone or in pairs at fruiting trees. The call is a high pitched *'psee-pseeee.'* Can be seen near the Machu Picchu Museum and on the grounds of the Machu Picchu Pueblo Hotel.

Red-crested Cotinga   *Ampelion rubrocristata*
21-21.5 cm. 2400-3700 m. Mostly plumbeous gray, blacker tail with a broad white band across the tail visible in flight. Semi-concealed crest of long maroon colored feathers laid flat over nape and usually difficult to see unless the bird is agitated or displaying when the crest is erected in spectacular fashion. The immature has streaked buff underparts and lacks crest. Found in humid montane and elfin forest, often at edges and in clearings. Usually conspicuous, perched upright at or near the top of tall trees and bushes. Sits still for long periods. Found singly or in pairs, this cotinga eats many fruits, especially mistletoe. The seldom heard call is typical of the genus – a short harsh, rasping frog-like *'grrrr.'* Found at elevations between 2600 m and 3600 m. Quite common at various localities along the Inca Trail and Abra Malaga.

Chestnut-crested Cotinga   *Ampelion rufaxilla*
21 cm. < 2700 m. A boldly patterned cotinga. Conspicuous elongated chestnut and black crest is usually laid flat across the nape. Eye bright red. Usually found perched at the top or close to the top of tall trees where it behaves much like the preceding species, sitting still for long periods. Feeds on fruits and sometimes fly-catches. The call is a harsh, seldom heard, frog-like *'ree-re-re-reh.'* Rare at Machu Picchu.

Andean Cock-of-the-rock   *Rupicola peruviana*
30-31 cm. < 2400 m. Large and unmistakable. The national bird of Peru. Both sexes show a bushy compressed crest which covers bill, and a bluish-white iris. Found in ravines and along streams in humid pre-montane forest, forest edge and mature secondary woodland. Moderately shy, often seen flying across clearings or rapidly down valleys. Sometimes congregate at fruiting trees. Best observed at display leks where up to a dozen males display daily in the early morning and late afternoon, bowing and jumping while flapping their wings and bill snapping. They also make a raucous croaking call constantly at the leks. Can be seen along the railway track between Puente Ruinas railway station and the Mandor Valley where there is an, unfortunately, inaccessible lek after about 1 km. Also seen, on the grounds of the Machu Picchu Pueblo Hotel.

BECARDS   Tityridae

Due to taxonomic uncertainty, this family which includes tityras, becards and allies had been tentatively placed in the cotingas. Formerly scattered among the flycatchers, cotingas and manakins; the Tityridae have now been elevated to family rank.

Barred Becard   *Pachyramphus versicolor*
12.5-13 cm. < 3000 m. The most frequently encountered becard at Machu Picchu. Note the chunky, hunchbacked shape. Found in humid montane forest and mature second growth woodland, sometimes at the edge. Usually encountered in pairs, often accompanying mixed feeding flocks where they are quite active. Perch and hover gleans from the top of leaves or branches for insects or small fruits. Common throughout the humid forests within the Machu Picchu Sanctuary. Often heard call is a high *'tu-wheeer-tu-tu-tu-tu.'*

Crested Becard   *Pachyramphus validus*
17.5-18.5 cm. < 3400 m. Large. Male: two toned dark gray above, light gray below and contrasting black cap. Female: mostly rufous with gray crown. The resident *audax* subspecies is found in small numbers in our area, particularly in side valleys that enter the main Urubamba valley from the Cordillera Vilcanota and the west side of Abra Malaga near Peñas. Often alone, but also with mixed species flocks. Found in semi-humid to dry shrubbery and riverside growth with alders. Make a variety of high squeaky notes. Uncommon.

VIREOS   Vireonidae

A strictly New World family. The vireos are a group of predominantly olive and gray arboreal birds. The sexes are similar. Vireos inhabit dense vegetation and shrubbery and are usually quite difficult to see, being more often recognized by song than seen. Mostly insectivorous, they will also eat small fruits. The nest is a finely-woven deep cup suspended from the crotch of a branch.

Rufous-browed Peppershrike   *Cyclarhis gujanensis*
15-16 cm. < 2500 m. The nominate form is present in our area. Gray head with bright rufous forehead and brow. Found in open humid montane forest, clearings, gardens, well-developed secondary growth and plantations. Rather sluggish, moving slowly in foliage, usually in pairs and sometimes with mixed species flocks. Forages inside the canopy and is quite difficult to see. The song is a series of rich and varied musical phrases, tirelessly repeated for minutes at a time and a descending *'cuyi-cuyi-cuyi-cuyi.'* Rare in our area.

Brown-capped Vireo   *Vireo leucophrys*
14-15 cm. < 2600 m. The *laetissimus* race is present at Machu Picchu. Olive above with a contrasting brown crown and prominent whitish supercilium. Likely to be encountered at the edges of humid montane forest, often in alders, gardens, lighter woodland and secondary growth. Forages deliberately at high levels and often accompanies mixed tanager flocks. The call is a rising buzzy *'zreee'* and the song is a short musical warble. Fairly common in the lower parts of the Machu Picchu Sanctuary.

Red-eyed Vireo   *Vireo olivaceous*
12-15 cm. The South American resident forms *(chivi)* may deserve full species rank, but more taxonomic work is needed. Olive above, gray crown, white superciliary and whitish below. Iris brown. The North American migrant subspecies *(olivaceous)*, may occur in small numbers between September and March. They are duller and grayer and the iris is red. Found in all kinds of humid forest except dense unbroken tracts, often in lighter woodland, humid forest-edge and clearings with scattered trees. Encountered alone or in pairs, often with mixed species flocks, perch gleaning with deliberate movements at mid to upper levels inside the canopy. Sometimes sallies or hangs. Mostly insectivorous during the breeding season, but feeds on fruits as well at other times of the year. The song of resident *'chivi'* is a series of short phrases repeated incessantly *'cheewewe-cheewe-cheeweewi-chirchirchir-cheeweewee.'* The call is a House Sparrow-like chirp *'tijrr.'* Can be seen at Machu Picchu along the railway track at the Puente Ruinas railway station.

JAYS   Corvidae

There are no true crows in South America and jays represent this family in the neotropics. They are colorful, unlike crows, omnivorous, and, for the most part, live in noisy family groups. Jays move actively through vegetation, searching all parts of the trees for food. The nest is a large open cup constructed of twigs.

White-collared Jay   *Cyanolyca viridicayanus*
30-32 cm. 2200 -3500 m. The *cyanolaemus* subspecies is present. Found in humid montane forest and forest borders. Groups of 3-8 birds move through vegetation at all levels, searching for food by hopping along and inspecting foliage, crevices etc. The species has a variety of calls quite loud and distinctive *'jeet-jeeet-jeet'* or *'jow-jow-jow.'* Despite the distinct call, size and color, they can be quite difficult to see in the forest. Can be seen along the Inca Trail between Phuyupatamarca and Intipunku.

Green Jay   *Cyanocorax yncas*
30-33 cm. < 2400 m. A most colorful jay, not likely to cause confusion. Found in humid montane forest, forest edge and borders, sometimes in lighter woodland. Like other members of the genus, they move around in family groups of up to 10 birds, conspicuously foraging at all levels. Twists tail continuously. Very vocal, giving a variety of calls including *'kerr-kerr-kerr'* and *'ya-nya-nya.'* Can be seen along the railway track between Puente Ruinas and the Mandor Valley at Machu Picchu.

SWALLOWS AND MARTINS   Hirundinidae

A familiar, cosmopolitan family, which superficially resembles swifts in that they are mostly aerial and catch prey on the wing, but are not as swift and agile as swifts. Unlike swifts, which can only cling to a surface, swallows and martins often perch on exposed twigs and phone wires. Nests are often made of mud and placed on cliff faces and also placed in holes and crevices. Most species are gregarious.

Brown-chested Martin   *Progne tapera*
17-18 cm 3000 m. Rare with one record from Huacarpay Lakes near Cusco. (austral migrant *fusca*?). Well defined dark breast-band. Glides with bowed wings, found singly or in pairs. Usually associated with marshes, river and lakes.

Purple Martin   *Progne subis*
18-19 cm < 3300 m. Boreal migrant. Male indistinguishable in the field from Southern Martin. Female lighter below unlike much darker Southern Martin. Rare in our area, has been seen repeatedly in passage at Huacarpay Lakes in October and likely at other locations.

Blue-and-white Swallow   *Pygochelidon cyanoleuca*
11-12 cm. < 3000 m. Adults are blue above and white below. Juveniles are duller above and dusky below and can be confused with other species. The tail is slightly forked. Widespread in a variety of habitats, along rivers, in open areas, near towns and agricultural fields. Usually in small groups hawking at all heights but mostly higher. Often seen perching in groups on exposed branches or wires. Sometimes perches on the ground. It is the common swallow at the Machu Picchu ruins, where it nests in the old Inca walls. Call a liquid *'chew.'*

Brown-bellied Swallow   *Orochelidon murina*
13.5 cm. 2500-4600 m. Note brown underparts. Found foraging over semi-humid forest, arid scrub, along cliff faces and even over open grassy areas (puna grassland). Usually in small groups but up to 50 birds or more can be found together, hawking at various heights. Flight call is a rasping *'tjirt-tjirt-tjirt.'* Common in the Cusco area.

Pale-footed Swallow   *Orochelidon flavipes*
11-12 cm. 2000-3500 m. Throat and upper breast pale pinkish buff (difficult to see in the field), brown flanks are a good field marks but easily confused with the species below. Usually found in groups of 5-20, hawking over humid montane and elfin forest and forest clearings. In general has a faster, more direct flight than the next species. Often found in mixed flocks with the Blue-and-white Swallow at lower elevations, and the Brown-bellied Swallow at higher elevations. Call is a buzzy mellow *'drzeee'* or *'jureeez,'* which, once learned, is useful for identification. Can be seen at Abra Malaga.

Andean Swallow   *Orochelidon andecola*
14 cm. 3500-4600 m. Pale below with darker throat. A rather thick-set swallow with triangular wings and a vaguely forked tail. Generally a high elevation bird found in small groups, sometimes alone, hawking over open country, puna grassland, rocky slopes and edges of small towns, sometimes around livestock. Nests in crevices in cliff faces, buildings and road cuts. Uncommon, but can be seen on the open grasslands at higher elevations. Call *'chirleep.'*

Southern Rough-winged Swallow   *Stelgidopteryx ruficollis*
13 cm. < 2000 m. Note contrasting pale rump. Found in small groups often associated with water, often perching on dead branches or wires. The outer web of the outer primary is indeed rough (serrated). Nests in holes in banks. Quite uncommon in our area. Has been recorded along the Urubamba River below Aguas Calientes.

White-rumped Swallow   *Tachycineta leucorrhoa*
13.5 cm 3000 -4000 m. Austral migrant. Note the narrow loral line. Rare migrant recorded at Huacarpay Lakes and Abra Malaga. To be looked for May to October.

Bank Swallow   *Riparia riparia*
12 cm. < 3500 m. A boreal migrant present in Peru from September to April. Has a distinct "fluttery" flight unlike the deep wing-strokes of other swallows they often accompany. Often with Barn Swallows. Usually seen in small to large groups over rivers, agricultural terrain and open hill slopes. The call is a buzzy *'drrrt.'* Most likely to be encountered near lakes around Cusco.

Barn Swallow   *Hirundo rustica*
14-16 cm. < 4000 m. Length depends on the size of the tail-streamers. A boreal migrant present in Peru from September to April. Tail deeply-forked, shorter in immature and migrating birds. In the Machu Picchu – Cusco area it inhabits open country, not likely to be found in heavily forested areas.

Cliff Swallow   *Petrochelidon pyrrhonota*
13-13.5 cm < 4000 m. Rare but regular boreal migrant. Prominent buffy whitish forehead and tawny rump. Could show up anywhere, has been seen at Lakes Piuray and Huacarpay and also at Machu Picchu ruins.

WRENS   Troglodytidae

A mainly New World family (only one species in the Old World). Most are a dull colored mix of rufous and browns with black barring on the wings and tail. They are renowned songsters, males and females often dueting. Most are furtive and live in dense undergrowth where they are mainly insect eaters. Only the female incubates, but both sexes feed the young. The nest is a round structure usually built by the male and placed in a cavity, bush or tree. Most species are highly territorial, vigorously defending their territories.

Gray-mantled Wren   *Odontorchilus branickii*
12 cm. < 2000 m. An atypical wren of the upper and middle stories of humid cloud forest, where it is often found with mixed feeding flocks. Actively gleans from side to side along mostly horizontal branches in a warbler-like manner, inspecting epiphytes and moss-covered branches, sometimes cocking its tail. The song is a thin, high-pitched, monotonous trill, given in the early morning or sometimes whilst accompanying flocks. The song is often overlooked. Uncommon at Machu Picchu and best looked for along the Urubamba River between Aguas Calientes and the Aobamba Valley. Named for Konstanty Branicki (1824-1884), a Polish aristocrat who employed many collectors in South America.

House Wren   *Troglodytes aedon*
12-12.5 cm. < 4600 m. A familiar wide-ranging bird. A small olivaceous – brown and buff wren with no particularly striking features, but often cocks tail. Found in a variety of open habitats, but absent from dense, closed forest. Often common around human habitation. Usually within 3 m of the ground but sometimes higher. Constantly active moving through low vegetation looking for fairly large invertebrates. Often tame. The frequently repeated song is a varied warble, rising and falling, lasting about four seconds.

Mountain Wren   *Troglodytes solstitialis*
10-11 cm. 2000-3600 m. Small, tawny with obvious buffy superciliary. The *macrourus* race is present at Machu Picchu and replaces the former species in humid forest. Forages at mid-levels, sometimes ranging into the canopy and often associated with viny tangles where it perch gleans for insects. Sometimes hitches along mossy limbs and branches investigating crevices for invertebrates. Can be found alone or in pairs but is also a prominent member of mixed feeding flocks of tanagers, warblers etc. Quite common in humid montane and sub-montane forest throughout the Machu Picchu Sanctuary.

Sedge Wren   *Cistothorus platensis*
10-11 cm. 3500-4000 m. The race at Machu Picchu may deserve full species rank; the name Grass Wren has been suggested. Usually found at tree line in humid situations such as boggy meadows with short bushes, or at the edge of very wet elfin forest. Does not like heavy forest and sometimes wanders out into quite open grassy habitats. Shy and difficult to see as it sneaks through bunch grass, rushes and the bottom of bushes, but easily identified by its streaked mantle. When flushed, it spreads its tail and flies low over vegetation. Can be found singly or in pairs along the Inca Trail between Sayacmarca and Phuyupatamarca and at Abra Malaga.

Gray-mantled Wren

House Wren

Mountain Wren

Sedge Wren

Inca Wren   *Pheugopedius eisenmanni*
14.5- 15.5 cm. 1800-3400 m. Peruvian endemic. A distinctively handsome wren. Found in pairs or small family groups, never with mixed feeding flocks, at the edge of humid forest where large stands of *Chusquea* bamboo are present. They probe dead leaf clusters and bamboo stem clusters. The song is a musical, loud series of notes – *'tui-wee-weet weee weh,'* repeated constantly. This song is one of the characteristic sounds of the Machu Picchu ruins, with the male and female singing antiphonally. Though difficult to locate, the bird can be found wherever there are stands of bamboo in the ruins. Only described to science in 1985 and restricted to the Machu Picchu area and vicinity. Named for Eugene Eisenmann (1906-1981) US/Panamanian lawyer and ornithologist.

Fulvous Wren   *Cinnycerthia fulva*
14-14.5 cm. 1800-3000 m. Found in small family groups of up to 12 individuals, in humid montane forest and borders, particularly where *Chusquea* bamboo is present. Never alone, and sometimes with mixed species flocks. This species moves quickly through the undergrowth, often close at hand, but difficult to see. Clambers up vertical stems and actively works through the undergrowth, inspecting moss and leaves for insects. The fine, loud song is a series of whistled phrases with a rich musical quality often given by more than one individual. Contact and alarm notes include *'trrr-trrr-trrr.'* Fairly common along the Inca Trail between Phuyupatmarca and Wiñay-Wayna ruins.

Gray-breasted Wood-Wren   *Henicorhina leucophrys*
11 cm. < 2800 m. A short-tailed wren of forest undergrowth. Inhabits the undergrowth of humid montane and sub-montane forest and mature secondary woodland, where it can be difficult to see. Always in pairs, not associated with mixed feeding flocks, creeping around in dense, dark undergrowth and tangles, investigating branches and stems for invertebrates. Very territorial and sings throughout the year, the song being a familiar sound of the forests of Machu Picchu. The song, often delivered by a duetting pair is a repetition of several long musical phrases – *'chee-wee-cheewee-wee-wee-churdlee-chee,'* etc. Common along the railway line between Aguas Calientes and the Mandor Valley.

DIPPERS   Cinclidae

Dippers are the only truly aquatic passerines. They are entirely adapted to rocky streams and show a number of adaptations as a result of this, including large feet, dense plumage, an unusually large oil-gland and a well developed nictitating membrane. They are invariably found in pairs and are strongly territorial. They build their nest on a ledge of a bank, often underneath a waterfall.

White-capped Dipper   *Cinclus leucocephalus*
15-15.5 cm. < 3500 m. Boldly patterned black and white and unmistakable. Inhabits rushing mountain streams (often quite small ones) and rivers with boulders both in forested and open areas. Usually seen perched out in the middle of a rushing torrent or along the bank, usually in closely associated pairs. They feed by picking insects from rocks and boulders at the water's edge and also infrequently submerge below the waterline in search of prey. In flight they fly fast and low over the water with rapid wing-beats. Very territorial. The song is a loud musical trill. Found at all elevations within the Sanctuary, but not much above 3500 m. Common along the Urubamba River and its tributaries.

Inca Wren

Fulvous Wren
with white on head

Gray-breasted Wood-Wren

White-capped Dipper

THRUSHES Turdidae

A cosmopolitan family of great diversity, most of the typical thrushes of South America have less colorful plumage than Old World species. Juvenile plumage is usually spotted. Most flick their wings and tail. True thrushes are partly terrestrial and move with strong hops, while solitaires are arboreal. They feed on invertebrates, worms, fruits and berries. Many species of the family are known for the beauty and versatility of their songs; males sing during the breeding season, mainly at dawn and dusk. The nest is an open cup in a tree or bush.

Andean Solitaire   *Myadestes ralloides*
16.5-18 cm. < 2800 m. Wings have silvery band across the base of the primaries and the dark brown tail has silvery-gray outer feathers, usually only visible in flight, but then conspicuous. The immature is rufous brown heavily spotted buff. Found in the lower middle strata of humid montane forest and forest edge. Quite inconspicuous except when attending a fruiting tree and more often heard than seen. Sits with an upright stance perch gleaning for fruits and berries. Also, fly-catches and foliage-gleans insects. Sings from a concealed perch and its song is one of the most well known of the forests, high pitched notes reminiscent of a badly oiled swinging gate – *'tlee-leedle-lee-lulee-treeleee,'* etc. Can be seen easily along the railway track between Aguas Calientes and the Mandor Valley at Machu Picchu.

Slaty-backed Nightingale Thrush   *Catharus fuscater*
17 cm. < 2900 m. Note dark head and red bill. Inhabits undergrowth of humid pre-montane forest where it is very shy and difficult to see. Sometimes encountered quietly hopping across the forest floor. Early in the morning, it occasionally ventures out into clearings and onto trails. More often heard than seen – the song consisting of melodic two or three note phrases. Only likely to be seen in the lower reaches of the MPHS, where it is rare.

Swainson's Thrush   *Catharus ustulatus*
16-18 cm. < 3300 m. Boreal migrant present from September to April. Shows buffy-white lores and eye-ring, giving a spectacled effect. Spotted breast. Found in montane and pre-montane forest and edges, secondary growth and adjacent clearings. Forages at low to mid heights, feeding on fruits and berries, often hovering and fluttering on the outside of canopies, where several will congregate at fruiting trees. Generally occurs below 2000 m but higher when migrating, such as gardens and peppertree groves in the Cusco area.

White-eared Solitaire   *Entomodestes leucotis*
22-23 cm. < 2900 m. Black and rufous with white band below eye. Inhabits humid montane and pre-montane forest at mid to high levels. Feeds on fruits mostly, remaining still for long periods hidden by foliage. Difficult to see unless encountered at a fruiting tree. The song is a strange nasal *'zeeeeee'* with an electronic quality, repeated every eight seconds or so and delivered from a concealed perch. Uncommon, but can be seen on the lower slopes of Machu Picchu mountain.

Pale-eyed Thrush   *Turdus leucops*
20-21 cm. < 2100 m. Male black with white iris. Female drab brown and easily confused with female Glossy-black Thrush. Found in the canopy and borders of humid montane forest. Usually encountered singly or in pairs, sometimes more at fruiting trees. Inconspicuous and not often seen. The song is a series of high, thin, sometimes musical phrases given at long intervals from a concealed perch. Can be seen along the railway track between Puente Ruinas railway station and the Mandor Valley at Machu Picchu in the early morning.

Black-billed Thrush   *Turdus ignobilis*
22-23 cm. A nondescript thrush of the lowlands at the limit of its elevational range at Machu Picchu. Note streaked throat and black bill. Sexes similar. Inhabits gardens, lighter woodland, clearings, plantations and semi-open areas. Not found in dense forest. Not shy, and can be encountered hopping around in pastures and clearings although it quickly retires to trees when disturbed. Shivers tail constantly. The song is typical of the genus, though perhaps less rich than most. Rare at Machu Picchu, recorded from the Mandor Valley.

Slaty Thrush   *Turdus nigriceps*
19-21 cm. < 2000 m. Austral migrant occurring May to October. Note streaked throat. An arboreal thrush of humid montane forest and borders. Usually encountered alone or in pairs at low to mid levels in dense vegetation along streams, sometimes in alders. Stays deep in vegetation and is difficult to see. Uncommon at Machu Picchu, seldom encountered.

Great Thrush   *Turdus fuscater*
30-33 cm. 2500-4200 m. The largest South American thrush, with the *ockendeni* race being present at Machu Picchu, which is much darker than the following species and shows a reddish bill. Found in humid montane and elfin forest and nearby clearings, where the Chiguanco Thrush is normally absent, but considerable overlap does occur. Can also be found in isolated, semi-humid forest patches in side valleys of the Urubamba River. Often seen hopping along the ground in fields and pastures and alongside trails where there is short grass. Often in fruiting trees. Roosts in loose flocks. Sings from a branch at mid-levels. The song is typical of the genus – varied whistled phrases. The alarm call is a loud *'kjee-ee-ee-ee-jee.'* Quite common at the tree line along the Inca Trail and at Abra Malaga.

Chiguanco Thrush   *Turdus chiguanco*   12/27 SACRED VALLEY, PURA
27-28 cm. 2400-4600 m. The nominate race is present in this area. Brownish with yellowish bill. One of the most familiar birds of the Andes. Found in mostly open country with bushes and shrubs, parks, gardens and agricultural land with hedgerows. Will colonize along roads through humid montane forest. Usually encountered alone or in pairs, hopping along the ground, searching for prey in short grass or along the edge of streams. Stops occasionally as if listening, with protruding breast and drooping wings. Sings from a high branch, telephone line or rooftop. The song is easily recognizable as a thrush consisting of 3-6 melodiously whistled phrases. Common.

Glossy-black Thrush   *Turdus serranus*
24-25 cm. < 3000 m. Smaller than the preceding species. Male black with orange-red orbital ring. Female similar to female Pale-eyed Thrush. Inhabits humid montane forest and forest edge though never ventures into clearings like the Great Thrush. Quite difficult to see, staying mostly at mid to high levels inside the canopy, where it feeds on fruits and berries. Sings from a high concealed perch. The song is repeated with intervals between phrases. Can be seen in the Mandor Valley and lower areas of the MPHS.

MOCKINGBIRDS Mimidae

Brown-backed Mockingbird *Mimus dorsalis*
24 cm. 3000 m. Vagrant from the south (Bolivia and NW Argentina). Only mockingbird recorded in the area so far but White-banded Mockingbird *(Mimus triurus)* could be expected. Feeds on the ground with cocked tail. Likes arid montane scrub; open bushy slopes with scattered cactus, agave and spiny plants. One record from Huacarpay Lakes.

PIPITS Motacillidae

Pipits are brown, streaked terrestrial birds found in grasslands. They are difficult to differentiate from one another. Pipits have slender bills and white outer tail feathers which help separate them from other streaked grassland species such as canasteros. In aerial display flight, the males fly to great heights and descend with vibrating raised wings and tail while singing. The nest is an open cup placed on the ground among bunch grass.

Short-billed Pipit *Anthus furcatus*
14-14.5 cm. 3500-4600 m. Breast and sides streaked dusky, breast band sharply demarcated from the white belly, giving a pectoral band effect. Found in grasslands and fields, seems to prefer short grass and rocky slopes and is found in drier locations than other pipits. Usually in pairs but also in loose flocks outside the breeding season. Terrestrial. In nuptial song flight, the male sings incessantly for up to 30 minutes before returning to the ground – *'trzseeuu-teedelee-leet-la-lee'* or variation. Can be seen around Lake Piuray and the highlands above Cusco.

Correndera Pipit *Anthus correndera*
14-15 cm. 3300-4500 m. The most heavily-streaked pipit in the area with prominent longitudinal "braces" on the mantle, the *calcaratus* race occurring here. Found in damp areas and adjacent dry puna grassland above tree line. Often in pairs or small groups, walking on the ground while feeding. Sits tight but flies long distances when flushed. The song given in the nuptial display flight or from the top of a small bush or tussock is a pleasant simple repetition of phrases – *'tzi-teer-gleeer,'* etc. repeated four or five times.

Paramo Pipit *Anthus bogotensis*
Length 15 cm. 3000-4500 m. Bill relatively long and stout. Rather sparsely streaked below and quite buffy. Found in pastures, fields, puna grassland and bogs. Behavior similar to the preceding two species, but display flight usually not as high. Will also sing from a bush or tussock – *'sweet-sweet-swzee-zee-zee-zee-eeeeeee.'*

Brown-backed Mockingbird

Short-billed Pipit

Correndera Pipit

Paramo Pipit

TANAGERS AND ALLIES   Thraupidae

A large diverse family, undergoing constant taxonomic change. A New World family, the majority of which are brightly colored with finch-like or, sometimes, slender bills. Most tanagers live in humid forest and feed on fruits and insects. A few species inhabit more arid regions. Many species of tanagers can be seen together in the same flock, some forage in the understory such as *Hemispingus* and *Thlypopsis* tanagers, but most, such as the *Tangaras*, feed in the middle to upper story. Only the female incubates, but both sexes feed the young. Nests are an open cup placed in a tree or bush. Also included: the small, warbler-like conebills, mostly bluish in coloration, with pointed bills, found in semi-open habitats and cloud forests; flowerpiercers with a distinctive upturned, hooked bill, which is used to pierce the base of flowers to extract nectar, colors are subdued, found mostly in forest borders and shrubby areas; and finches, most have relatively heavy conical bills adapted to opening seeds, but many species will take insects as well, found in open and semi-open country.

Magpie Tanager   *Cissopis leveriana*
25-29 cm. < 2000 m. An unmistakable large tanager with a very long tail. This large black and white tanager is a bird of clearings, lighter woodland, forest borders and river edge. Usually alone or in small groups and does not accompany mixed flocks as a rule, but may join aggregations of frugivores at fruiting trees. Usually feeds below 12 m in trees. The call is a loud, harsh *'chek.'* The song is explosive and short, with low-pitched and scratchy rattles *'t-t-t-t-t-ttreek-treek-treeek titl-titl-tleechuk-chuk.'* This mainly Amazonian species is rare at Machu Picchu, normally only occurring to 1400 m in Peru, but it has been recorded near the Mandor Valley at 2000 m.

Slaty Tanager   *Creurgops dentate*
14 cm. < 2500 m. Note the difference in plumages between the male and female. Found in humid pre-montane forest and well-developed secondary growth. Usually travels in pairs as part of a mixed species flock of *Tangara* tanagers. Mostly forages in the canopy and sub-canopy, well above the ground, peering deliberately in search of insects along twigs and limbs, usually in the outermost foliage. Song high pitched *'wheedling'* notes. Fairly common at Machu Picchu near Aguas Calientes. Can be seen on the grounds of the El Pueblo Hotel.

Black-capped (White-browed) Hemispingus   *Hemispingus atropileus auricularis*
15 cm. 2300-3300 m. The *auricularis* race occurs in our area and may deserve species status and, if so, would become a Peruvian endemic. Note white superciliary contrasting with black crown and sides of face. Inhabits humid montane and elfin forest, often where there are stands of bamboo (*Chusquea* sp.) This species moves in groups of 3-7 individuals, sometimes in pairs, and often with mixed feeding flocks which include Citrine Warbler. Moving fast and low through dense vegetation, usually below eye level, they creep up the stems of bamboo and other plants, often perch gleaning on the undersides of leaves and poking into bamboo internodes. The call is a harsh *'zit.'* The song is a high-pitched jumble of squeaky, sputtering notes. Common at Abra Malaga.

Parodi's Hemispingus   *Hemispingus parodii*
14 cm. 2600-3500 m. Peruvian endemic. Very similar to Citrine Warbler and often found in the same mixed species flocks. Note the stout tanager-like bill which is not black. Found in bamboo (*Chusquea* sp.) thickets in humid montane and elfin forest near the tree line. Typically, groups of 3-10 birds follow mixed feeding flocks, foraging low in bamboo and other admixed foliage, perch gleaning leaves and bamboo internodes. The song is a moderately pitched chittering *'p-p-p-p-psit-zit-zit'* lasting 1-2 seconds. Best looked for in bamboo stands at Abra Malaga. Named for José Parodi Vargas (c. 1930-1974), Peruvian politician and landowner.

Superciliaried Hemispingus   *Hemispingus superciliaris*
13-14 cm. 2200-3420 m. The *urubambae* race is present at Machu Picchu. White broken eye-ring. Upperparts olive, underparts yellow. Inhabits humid montane forest and mature second growth, sometimes in quite light woodland. Mostly in pairs or small family groups accompanying mixed species flocks at mid to high levels. Likes alders. The call is a high-pitched *'psit.'* The song is 3-4 seconds of dry harsh notes which increase in volume as they accelerate. Quite common.

Oleaginous Hemispingus   *Hemispingus frontalis*
14 cm. < 2600 m. Generally, a bird with few distinctive marks. Note the heavy bill. Inhabits humid montane and sub-montane forest in patches of bushes, dense thickets and vine tangles. Travels in small groups of 3-5 birds at mid-levels and in the undergrowth, twitching and flicking its tail, often with mixed species flocks. Feeds by perch gleaning leaves for invertebrates, sometimes hanging on leaves at the end of branches. Also, investigates dead leaf clusters in which it seems to specialize. The song is a series of high-pitched squeaky rattling notes *'chip-chip chee-chee ch-ch-ch-ch-wa-chew-wa-chew.'* Call *'chip.'* Can be seen in dense foliage near Puente Ruinas railway station at Machu Picchu.

Black-eared Hemispingus   *Hemispingus melanotis*
13-14 cm. < 2200 m. The *berlepschi* race is present at Machu Picchu. Note black face and tawny breast. Inhabits undergrowth with a predominance of bamboo inside humid montane and pre-montane forest and forest edge. Travels in groups of 3-7, sometimes alone or in pairs, often with understory mixed feeding flocks. Keeps quite low, perch gleans foliage and hangs from leaves searching for invertebrates. Will eat small berries. Call is a squeaky *'ja-sit.'* The song is a rapid twitter.

Drab Hemispingus   *Hemispingus xanthophthalmus*
Length 13 cm. 2400-3500 m. Quite a thin bill. Prominent whitish iris. Found in humid montane forest and forest edge. Also, occasionally moves into elfin forest. Forages in the upper foliage of tall bushes and trees from eye level to the upper canopy. Feeds and often perches on the tops of closely-packed stiff leaves. Small groups of 2-5 birds move with mixed species flocks (probably a nuclear species). The song is a pattern-less series of spluttering notes.

Three-striped Hemispingus   *Hemispingus trifasciatus*
13 cm. 3000-3400 m. Thin bill. Head black, with long buffy white supercilium. Ochraceous below. Mostly found at or near the tree line in low forest, elfin forest and low parts of humid montane forest. Groups of 3-7 individuals travel with mixed species flocks (appears to be a nuclear species) or on their own. They forage by perch gleaning in dense foliage in the crowns of trees or bushes toward the end of limbs. Also they glean twigs and probe mossy branches. The song at dawn is an endlessly repeated *'zwit.'*

Gray-hooded Bush-Tanager   *Cnemoscopus rubrirostris*
14-15 cm. 2000-3000 m The dark billed *chrysogaster* race is present at Machu Picchu. Fairly common in humid montane forest and forest borders. Actively forages along limbs and in foliage at middle and upper levels, often with mixed flocks of other tanagers. Wags its tail often and sometimes the whole rear of the tail seems to be moving. Calls include *'tsip'* and *'sstit.'* Can be seen along the Inca Trail between Phuyupatamarca and Wiñay Wayna.

Rufous-chested Tanager   *Thlypopsis ornata*
12 cm. 2000-3400 m. Note white belly contrasting with rufous underparts and head. At the southern end of its range at Machu Picchu. Inhabits dense shrubbery, forest edge, secondary growth and open woodland. Stays away from dense humid forests. Individuals, pairs or small groups of 3-5 travel alone or sometimes with mixed feeding flocks. Forages quickly in dense bushes, usually low, but as high as 6 m. Perch gleans twigs, leaves, flowers and dead leaf clusters. Call note *'seep.'* Uncommon.

Rust-and-yellow Tanager   *Thlypopsis ruficeps*
12 cm. 2000-3400 m. Crown and sides of head tawny, underparts yellowish. Inhabits bushes, gardens, secondary growth, bamboo and humid forest edge. Single birds, pairs or small family groups forage at low to medium heights, alone or with mixed species flocks. Sometimes ascends to the crowns of low trees. Feeds by perch gleaning the undersides of leaves, flowers and bamboo leaves and internodes. The song, often given by a pair, is a series of chittering notes, lasting about 1.5 seconds. Common at Machu Picchu and can be seen around the Machu Picchu ruin complex and along the road from the ruins to Aguas Calientes.

White-lined Tanager *Tachyphonus rufus*
17 cm. < 2000 m. An isolated population of this widely distributed species occurs in the Urubamba Valley and at Machu Picchu. Bill has a silvery blue sheen. Male appears all black. Female rufous. Shows white underwing coverts in flight usually not visible when the bird is at rest. Inhabits thickets, open woodland, gardens, secondary growth and shrubbery. Usually forages low to the ground in pairs, but will ascend to higher trees for fruit. Can be difficult to observe. Calls include a *'chek'* and the song is reported to be the repetition of a single phrase *'cheep-choi cheep-choi,'* etc. Has been recorded at 2000 m at Machu Picchu, along the railway line between Puente Ruinas railway station and the Mandor Valley. Local and uncommon.

Silver-beaked Tanager *Ramphocelus carbo*
16-17 cm. < 2200 m. Male: swollen bill, gleaming silvery white. Inhabits shrubby clearings, gardens and forest borders. Forages in noisy groups or pairs, which move rapidly through the undergrowth and shrubbery, pausing on occasion to eat fruit or glean leaves for insects. Usually not with mixed species flocks. The call is a harsh *'chink.'* At the upper limit of its altitudinal range at Machu Picchu and not encountered in large groups as in the Amazonian lowlands. Can be seen in the Mandor Valley. Uncommon.

Hooded Mountain-Tanager *Buthraupis montana*
23-24 cm. 2300-3500 m. A very large tanager. Note the obvious red iris, black hood, blue upperparts and yellow underparts. Inhabits humid montane forest and forest edge as well as tall secondary growth. Typically travels in groups of 3-10, alone or with mixed feeding flocks, often with Southern Mountain Cacique and White-collared Jay. Active and wide ranging, often flying long distances from one tree to another. Forages mostly on moss-covered limbs in the sub-canopy of tall trees. Quietly eats fruits, or searches sluggishly for insects in moss or epiphytes. The calls are a series of squealing and nasal *'week'* and *'toot'* notes, repeated rapidly. Display flight calls are high-pitched *'seet'* notes. In display, flies high on deep wing-beats before diving into a tree crown. Can be seen along the Inca Trail and at Abra Malaga.

Grass-green Tanager *Chlorornis riefferii*
20 cm. 2200-3000 m. A distinctive tanager of the humid forest. Bright green with a chestnut red mask. Found in wet, mossy humid montane forest and edge, less commonly in tall secondary forest. Usually encountered in groups of 2-6 individuals, with or separate from, mixed feeding flocks. Perches for long periods. Moves sluggishly from branch to branch when feeding, usually in the upper half of small trees. The song is a short phrase *'chink-zeet'* every two seconds. Call is a nasal *'enk.'* Fairly common at Machu Picchu along the Inca Trail. Named for Gabriel Rieffer who collected birds (1830s) in South America.

Lacrimose Mountain-Tanager *Anisognathus lacrymosus*
16-17 cm. < 3500. At the southern end of its range at Machu Picchu. Small spot below the eye (teardrop) yellow. Inhabits humid montane forest and elfin forest. Found mostly in pairs, sometimes in small groups, alone or with mixed species flocks. Moves widely between foraging sites. Feeds in tall bushes and small trees, ascending into the sub-canopy of larger trees. Searches quietly for berries and fruit at the ends of branches, leaning forward to pluck them. Hops and peers for insects, working its way from low in the center of a bush to the upper outside foliage. The song is a series of high-pitched squeaky notes given in 2-6 phrases – *'chuck-zit-zit swit-ee-dit swick ee-deee-deee,'* etc. Uncommon.

Scarlet-bellied Mountain-Tanager *Anisognathus igniventris*
16 cm. 2600-3600 m. Orange-red crescent behind the ear coverts with blue shoulder patch. Inhabits humid montane forest, elfin forest, forest edge and thickets at tree line. Encountered in pairs or small family groups of 3-8 individuals, alone or accompanying mixed species flocks. This species forages at all levels but mostly at mid-levels. Quickly moves along the edge or outer foliage of the forest. Plucks berries from trees by leaning out from a perch. Does take some insects. The call, also given in flight, is a high-pitched *'pst-tit.'* The true song is a complex series of twittering notes on a moderate pitch. One of the more common tanagers near the tree line.

Blue-winged Mountain-Tanager   *Anisognathus somptuosus*

16-17 cm. < 2300 m. The nominate race (Lesson, 1831) is present at Machu Picchu. Found in humid sub-montane forest, forest edge and tall secondary growth. Found alone or in pairs or in small family groups of 3-10, sometimes with mixed species flocks of other tanagers. Active, and forages rapidly in the canopy and sub-canopy, sometimes descending lower at the forest edge. Eats berries from a perched position leaning forward, sometimes hanging. Will take insects, which it finds by inspecting moss clumps, foliage and branch surfaces. Call is a thin *'tic'* and a series of these makes up the song at Machu Picchu. Best looked for along the Urubamba River between Aguas Calientes and the Mandor Valley.

Buff-breasted Mountain-Tanager   *Dubusia taeniata*

20 cm. 2300-3400 m. The *stictocephala* race is at Machu Picchu. A large, potbellied tanager. Found in low, dense vegetation, including bamboo thickets, forest edge and dense tangles inside humid forest. Sometimes feeds close to the ground in ferns at the tree line. Single birds or pairs may be encountered with mixed species flocks. Typically forages within thickets and shrubs and the foliage of small trees, usually keeping hidden. Sometimes sits atop a small bush to look around. Hops along mossy limbs, scanning the upper surfaces and bending over to inspect the undersides. Also, probes mossy clumps and lichens. The song is a high-pitched *'peeouee peeouee paaaay.'* Fairly common at Machu Picchu and Abra Malaga.

Chestnut-bellied Mountain-Tanager   *Dubusia castaneoventris*

15 cm. 2200-3500 m. Bluish above, chestnut below. Inhabits humid montane and elfin forest, forest patches and edge near the tree line. Usually encountered singly or in pairs, sometimes with mixed feeding flocks. Moves fairly slowly within trees but remains constantly on the move to keep pace with flocks. Usually forages just below the crowns of tall trees, occasionally lower. Picks berries while perched. Searches for insects while deliberately moving along moss-covered branches and limbs. Peers at moss, leaves and bromeliads. The song is a series of moderately pitched whistles *'peeee-pay-aay,'* the first higher-pitched than the second. Can be seen along the Inca Trail and at Abra Malaga.

Yellow-throated Tanager   *Iridosornis analis*

15 cm. < 2300 m. Startling yellow throat and rusty undertail coverts. Inhabits humid sub-montane forest with thick undergrowth. Sometimes at forest edge. Pairs and small groups of 3-6 follow mixed species flocks and join aggregations at fruiting trees. Feeds on fruit and insects, searching for them on lower branches and at mid-heights. The call is a descending *'tseeeer.'* Fairly common along the Urubamba River downstream from Aguas Calientes.

Golden-collared Tanager   *Iridosornis jelskii*

14 cm. 3000-3600 m. Note chestnut underparts. Found in low trees and bushes, often bamboo, near the tree line and forest edge, often in elfin forest and isolated forest patches above tree line. Can be encountered with mixed species flocks, but equally likely to be found alone or in pairs. Forages from low to mid-levels. Perches upright and leans down or sometimes hovers in order to pluck berries. Also, searches for insects on thin branches and among leaves at the end of branches. Calls are high-pitched – a series of *'see-see'* notes. Can be seen along the Inca Trail and at Abra Malaga.

Yellow-scarfed Tanager   *Iridosornis reinhardti*

14 cm. 2100-3000 m. Peruvian endemic. Head and upper mantle black with broad, yellow nuchal collar. Underparts blue. Found in humid montane forest shrubbery and edge. Usually in pairs or small groups, may accompany mixed species flocks. Moves through the interior of vegetation, hopping from branch to branch, at low to mid heights, sometimes in the canopy of small trees. Picks berries from a perch and hops along moss-covered branches looking for invertebrates. Will hop on the ground, probing moss clumps and tree roots. Call is a thin *'tzit.'* Found usually below the elevational range of the preceding species. Uncommon at Machu Picchu and here at the extreme southern limit of its range.

Fawn-breasted Tanager   *Pipraeidae melanonota*
14 cm. < 3000 m. Black face mask, bright red iris. At Machu Picchu, this species inhabits semi-open habitats, secondary growth, gardens and lighter woodland. Sometimes in quite dry habitat. Found singly or in pairs, usually not with mixed species flocks but will join feeding aggregations at fruiting trees. Forages at all levels when not in the forest, and in the canopy when inside forest or woodlands. Picks insects and berries among leaves at the tips of branches and sometimes sallies for insects. Fairly common at Machu Picchu near Puente Ruinas.

Blue-and-yellow Tanager   *Pipraeidae bonariensis*
16.5-17 cm. < 4000 m. The *darwinii* race is present at Machu Picchu. Note yellow rump in flight. Found in a variety of habitats, but prefers drier habitats than other tanagers. Can be seen in humid temperate forest but is more common in dry scrubby hillsides with cactus, open areas, orchards, gardens and agricultural areas. Usually found singly or in pairs away from mixed species flocks. Forages at low to medium heights, eating fruits including cultivated varieties where this species does much damage. Often seen at cactus fruits. Common around Huacarpay Lakes.

Orange-eared Tanager   *Chlorochrysa calliparaea*
12 cm. < 2200 m. The *fulgentissima* race is present. Above mostly bright shining emerald green. Note the burnt orange patch on the side of the neck. Found in humid mossy sub-montane forest. Pairs and single birds are usually conspicuous members of mixed species flocks. Forages mostly in the canopy and sub-canopy, hopping along, clinging to, or hanging from mossy branches and trunks, probing moss for invertebrates. Hangs or leans out to perch glean foliage. Also eats small berries. Call is a high-pitched wheezy *'seep.'* Fairly common at Machu Picchu along the Urubamba River between Aguas Calientes and the Mandor Valley.

Blue-gray Tanager   *Thraupis episcopus*
16 cm. < 2200 m. In the race at Machu Picchu, the shoulder and upper wing coverts are white. Found in a variety of habitats such as forest borders, secondary woodland, gardens and agricultural areas. Very active and often tame, always in trees and feeds at all levels. Perch gleans for insects and sometimes fly-catches. Also eats fruit. The song is a jumbled series of squeaky fast notes and the calls are similar, if somewhat shorter. Can be seen easily around Aguas Calientes.

Sayaca Tanager   *Thraupis sayaca*
16-17.5 cm. 3300 m. Vagrant, probably austral. Extremely similar to the immature of the preceding species and great care must be taken in identifying this species, even in the hand. Note white undertail coverts. One bird coming to feeder in Cusco was probably the *bolivianus* race of this species.

Palm Tanager   *Thraupis palmarum*
17 cm. < 2200 m. Wing coverts chalky olive, contrasting with the black terminal half of the closed wing. Inhabits shrubby clearings, gardens, and agricultural areas, as well as the forest canopy in sub-montane forest. As its name suggests, it likes palm trees. Forages for insects usually in pairs, often hanging upside down at quite high levels. Song and calls similar to the Blue-gray Tanager, with which it often associates. Rare at Machu Picchu.

Blue-capped Tanager   *Thraupis cyanocephala*
16 cm. 2000-3000 m. The *cyanocephala* race is present at Machu Picchu. Flanks and vent olive-yellow, thighs and under wing, bright yellow. Inhabits humid montane forest, usually in broken canopy and edge, as well as clearings and second growth. More numerous in areas with alders (*Alnus* sp.). Found alone or in pairs, rarely with mixed species flocks. Forages in the top and center of trees and tall bushes, searching large branches for prey. Sometimes hovers. Usually restless and conspicuous. The call is a high-pitched *'zit.'* The song is a series of short, squeaky notes, repeated 3-5 times. Common at Machu Picchu.

Golden-naped Tanager   *Tangara ruficervix*
13 cm. < 2400 m. Distinctive nape patch cinnamon-rufous. Buff belly. Inhabits humid sub-montane forest, near clearings, tree falls or isolated tall trees and forest edge. Found singly or in pairs, sometimes in small groups, mostly with mixed species flocks of other tanagers. Forages at mid to upper levels in the crowns of trees and tall shrubs, often in flowering and fruiting trees. Picks berries and probes at catkins of *Cecropia*. Sallies into the air and searches in the canopy for insects. Calls *'tsip.'* Can be seen at Machu Picchu on the railway track near Puente Ruinas railway station.

Silvery Tanager   *Tangara viridicollis*
13 cm. < 2700 m. The nominate race is present at Machu Picchu. Note coppery throat. Inhabits humid sub-montane forest edge, secondary growth, gardens and wooded ravines. Found in pairs or groups of 3-6, with or away from mixed feeding flocks. Forages from the tops of small trees to low bushes and in the humid scrub on hillsides. Eats berries and searches for insects on limbs with a horizontal stance. Call is a thin *'tziu'* and *'peew.'* Fairly common at Machu Picchu, especially around Puente Ruinas railway station.

Blue-necked Tanager   *Tangara cyanicollis*
12 cm. < 2400 m. A stunning mostly black tanager with contrasting blue head and golden wing coverts. Inhabits forest edge and open clearings, tall secondary growth and gardens close too humid sub-montane forest. Usually found alone, in pairs or small family groups, not usually with mixed species flocks, but will follow them occasionally at the forest edge. Joins feeding aggregations at fruiting trees. Forages mostly in the crowns of bushes and trees. Hangs from leaves for berries, inspects lichen-covered branches and also bare limbs, and sallies for insects. Inspects catkins of *Cecropia* and flower heads. Calls include *'chep,'* *'zeet'* and *'seep.'* Pleasantly common at Machu Picchu in the grounds of the El Pueblo Hotel in Aguas Calientes.

Blue-and-black Tanager   *Tangara vassorii*
13 cm. 2400-3500 m. The *atrocerulea* race is present at Machu Picchu. Flight and tail feathers black and edged with blue. Inhabits humid montane and elfin forest, forest edge, tall secondary growth and clearings. In pairs or family groups of 3-6, mostly with mixed feeding flocks but also attends feeding aggregations at fruiting trees. Forages at mid to high levels, staying in the crown when inside the forest. Perches on branches and leans down to pluck berries. Searches for insects while hopping along slender moss-covered branches. Probes mossy clumps and perch-gleans leaves. The call is an emphatic *'zwit'* or *'swit swit.'* Generally replaces the previous species at higher elevations. Can be seen along the Inca Trail and at Abra Malaga.

Beryl-spangled Tanager   *Tangara nigroviridis*
12 cm. < 2400 m. Rump turquoise blue. Heavily scalloped greenish-blue below. Inhabits humid montane and sub-montane forest, forest edge, tall secondary growth and tall trees in clearings. Travels in pairs or groups of up to 10. Often a conspicuous member of mixed species flocks and at aggregations at fruiting trees. Feeds at mid to high levels. Peers at leaves, gleans bare twigs and pokes hanging moss clumps. Leans sideways, up and down, stretching out from its perch for berries or insects. Can be seen along the Urubamba River between Aguas Calientes and the Mandor Valley. Beryl refers to a mineral occurring in green, bluish green, yellow, pink, or white hexagonal prisms.

Bay-headed Tanager   *Tangara gyrola*
13-14 cm. < 2000 m. The *catherinae* subspecies is present. Greenish-blue with contrasting chestnut head and yellow nuchal band. Forages for insects by perching on mainly open branches in humid sub-montane forest, often tilting and peering on the undersides. Rare at Machu Picchu but has been seen along the railway track near Aguas Calientes.

Saffron-crowned Tanager *Tangara xanthocephala*
Length 13 cm. < 2400 m. The *lamprotis* race is present at Machu Picchu. Crown orange (not saffron). Inhabits humid montane and sub-montane forest, forest edge and tall secondary growth. Also scattered trees in open areas near forest. Pairs and groups of 3-10 accompany mixed species flocks. Restlessly forages in the outer branches of trees for fruits and insects, often hanging and peering under branches. One of the most frugivorous of tanagers. The call is a crisp *'zsit.'* One of the more familiar tanagers at Machu Picchu.

Flame-faced Tanager *Tangara parzudakii*
14-15 cm. < 2600 m. A gaudy tanager at the southern end of its range at Machu Picchu. The *urubambae* race is present. Inhabits humid montane and sub-montane forest and forest edge. Pairs and groups of 3-7 individuals travel with or away from mixed species flocks. Sometimes perches quietly in tree tops for long periods. Forages for small fruits and insects in the upper levels of the forest, sometimes lower at the forest edge. Perches while feeding on berries, sometimes hanging upside down to pluck them. Searches slowly and deliberately for insects on mossy branches and limbs, sometimes probing mossy clumps. Call is a *'chit'* sometimes given in series (song?). Can be seen along the Urubamba River between Aguas Calientes and the Mandor Valley. Named for Charles Parzudaki, French traveler and collector in Colombia in the mid 1800's.

Blue Dacnis *Dacnis cayana*
11-12 cm. < 2200 m. Note the long pointed bill. Legs pale flesh. At Machu Picchu inhabits forest edge and secondary growth, gardens, and semi-open areas close to the sub-montane forest. Alone or in pairs, often with mixed species flocks or aggregations at fruiting trees. Forages high in trees taking insects with acrobatic motions. Perches and hangs upside down on twigs and also sallies for insects. Perch gleans both sides of leaves. Eats small fruits and takes nectar. The call is a high pitched *'zit'* or *'chit.'* Occurs along the Urubamba River near Aguas Calientes.

Cinereous Conebill *Conirostrum cinereum*
11-12 cm. 2500-4000 m. The pale wing patch and white supercilium are good identification features. Mostly inhabits semi-arid habitat, gardens, agricultural areas, semi-open areas with scattered bushes and trees, riverine thickets etc. Tends to avoid humid forest, but sometimes found at the tree line in elfin and *Polylepis* forest. Usually in pairs or small family groups, restlessly flitting between small trees and bushes. Feeds mostly on the inside of vegetation, investigating branches for insects and small berries. Often clings upside down while searching for food. Sometimes with mixed flocks. Song a quick assortment of high squeaks and chatters. Call *'chip.'* Common.

Blue-backed Conebill *Conirostrum sitticolor*
12-13 cm. 2500-3800 m. Shows glossy purplish blue supercilium. A distinctive conebill found in humid montane and elfin forest and forest borders. Usually in pairs or small groups, often accompanying mixed feeding flocks, foraging actively at the tips and outsides of bushy foliage, frequently hanging upside down. Sometimes sallies for insects. Common at Abra Malaga.

Capped Conebill *Conirostrum albifrons*
12-13 cm. < 2800 m. The *sordidum* race is present at Machu Picchu. Male can appear all black. Slender pointed bill. Found in humid montane and sub-montane forest canopy and high second growth, mostly with mixed feeding flocks, but sometimes alone or in pairs, usually foraging quite high along branches and stems, actively gleaning and probing. Fairly common along the Urubamba River between Aguas Calientes and the Mandor Valley.

White-browed Conebill *Conirostrum ferrugineiventre*
12-12.5 cm. 3000-4000 m. Note chestnut underparts and obvious white superciliary. Inhabits bushy areas at the edge of humid montane, elfin and *Polylepis* forest. Usually encountered in pairs or small family groups, often accompanying mixed feeding flocks of tanagers and warblers. Song a high pitched jumble of notes, call *'tzit.'* Common along the Inca Trail and at Abra Malaga.

Giant Conebill   *Oreomanes fraseri*
14-16 cm. 3300-4500 m. A large conebill with a heavy pointed beak. Much variation in intensity of color in the same population. Restricted to *Polylepis* forest at high altitude. Mostly found in pairs or family parties on medium to thick branches, less often on smaller twigs, mostly feeding inside the canopy. Typically probes under flaking *Polylepis* bark, often flaking off large pieces when searching for invertebrates. Hitches along trunks and branches, sometimes hanging upside down. Visits *Gynoxys* plants for *aphids*. The pleasant song is a series of musical, repeated phrases *'chhet-chip-cheveet-chjeeveet,'* etc. Easily seen in *Polylepis* at Abra Malaga. Named for Louis Fraser (1819-1883?) British zoologist and diplomat.

Tit-like Dacnis   *Xenodacnis parina*
12 cm. 3200-4500 m. The *parina* race is present in our area. Inhabits patches of woody plants at or above the tree line, including forest edge and *Polylepis* woodland. Confined to areas where shrubs of *Gynoxys* are present where it gleans the undersides of leaves in a very active manner looking for *aphids* and sugary secretions (crystal and droplets). Usually encountered alone or in pairs, sometimes with mixed species flocks of warblers and flowerpiercers. Forages in low shrubs and small trees, sometimes in the crowns of medium sized trees. Calls include a soft *'tchiu-tchiu-tchiu'* and *'zweet-zweet-zweet,'* not as loud as birds further to the north. At Abra Malaga and even above Cusco.

Moustached Flowerpiercer   *Diglossa mystacalis*
14 cm. 2500-3600 m. The *albilinea* subspecies is present at Machu Picchu. Sexes similar. Distinctive off-white moustache and chestnut abdomen and vent. Blue-gray shoulder patch. Inhabits humid montane forest edge, elfin forest and patches of thickets surrounded by puna grassland. Sometimes in *Polylepis* woodland. Usually solitary and not found with mixed species flocks. Stays hidden in vegetation. Very aggressive and territorial, often driving away conspecifics as well as other species. Forages close to the ground in shrubbery, gleaning the undersides of small leaves and stems by stretching and reaching from a perch. Obtains nectar by piercing the base of flowers. The song is a loud and not unpleasant twittering *'twee-t-deetdee-teerwheet-weetee,'* etc. Quite common on stretches of the Inca Trail and at Abra Malaga.

Black-throated Flowerpiercer   *Diglossa brunneiventris*
13 cm. 2000-3800 m. Found in rather dry areas but also sometimes in more humid situations. Inhabits scrub, watered ravines, gardens and agricultural areas. Usually encountered alone or in pairs. Very active and constantly on the move at low to medium levels, sometimes in tree tops where *Eucalyptus* are present. Searches for nectar and insects among flowers and at the tips of branches. Both pierces and directly enters flowers. Gleans leaves inside dense vegetation, often hanging upside down. Song is a warble lasting 2-4 seconds *'tee trree ree reee ri tree.'* Call a high-pitched *'zsit.'* Common around Cusco.

Rusty Flowerpiercer   *Diglossa sittoides*
11 cm. < 2500 m. Inhabits open country with scattered bushes and trees, gardens, and cut-over slopes. Female can be confusing. Usually alone and in pairs away from mixed species flocks. Active and quick, often chased by hummingbirds. Forages by piercing for nectar and gleaning insects, mostly on flowering shrubs and trees at low to mid-levels. The call is a loud sharp *'cheek.'* Song is an unremarkable twitter. Can be seen on scrubby hillsides around the Machu Picchu ruins.

Bluish Flowerpiercer   *Diglossa caerulescens*
12.5-13.5 cm. < 2600 m. A uniform looking flowerpiercer – grayish blue, and a little paler below, especially on the belly. Inhabits humid montane and pre-montane forest and forest borders. Usually encountered alone or in pairs with mixed feeding flocks, including flocks containing Masked Flowerpiercer. Slow and inconspicuous, feeding at mid-levels, puncturing flower bases for nectar. Also, gleans leaves for insects and takes small fruits. Call is a loud *'chink.'* The song is a series of high-pitched melodic notes which run into a squeaky twitter lasting 2-3 seconds. Can be seen in the forest near Intipunku on the Inca Trail.

Masked Flowerpiercer   *Diglossa cyanea*
15 cm. < 3000 m. Bill long and black. Iris bright red. Inhabits humid montane and elfin forest and forest edge. Also, in nearby secondary growth and areas with scattered trees and bushes. Found in pairs or family groups and commonly found as part of mixed species flocks. Forages at different heights, from mid-story to the canopy, and even in bushes at the forest edge. Flits and hops around gleaning leaf surfaces. Reaches out and leans from a perch searching for prey. Eats fruit and also feeds on nectar by directly entering the flower. The song is a very high-pitched *'tzi tzi tzi tzi tziweedlideeleedeedleeee'* and variations on the theme. Call is a fine *'zit.'* Common at Machu Picchu and Abra Malaga in forested areas.

Buff-throated Saltator   *Saltator maximus*
20-21 cm. < 2200 m. Saltators are large, heavy billed arboreal finches with melodious whistled calls. Distinct white supercilium and black mustachial streak. A common Amazonian bird, rarely reaching Machu Picchu. Found in canopy and sub-canopy of humid forest and taller secondary growth, usually at edge and in clearings. Song a thrush-like complex song. Call a harsh metallic *'chiiik.'* Rare.

Golden-billed Saltator   *Saltator aurantiirostris*
20 cm. 2000-3700 m. Bill entirely or patchily bright golden orange – dark in immature. Conspicuous white post-ocular stripe. Outer tail feathers broadly tipped with white. Found in both humid and dry montane scrub and shrubbery, secondary growth, hedgerows and gardens, and stands of *Eucalyptus* trees, alone, in pairs or family groups. Forages for insects, seeds and berries on or near the ground. Quite shy, but sometimes perches in the open. Sings from a high, hidden perch in a tree. Song consists of 3-5 syllables *'dee dee te teuie tri te tuit trt trrrt tiuiet.'* Call is a loud *'chack.'* Common around Cusco.

Plushcap   *Catamblyrhynchus diadema*
13-14 cm. 2000-3600 m. Bill black, stubby and rounded. Golden yellow forehead. Inhabits bamboo and understory in humid montane and elfin forest. Found in pairs or family groups, often with mixed species flocks. Seems to be a bamboo (*Chusquea* sp.) specialist, feeding inside the vegetation, hanging from tips or stalks. Pecks and tugs at leaves and gleans bamboo stems with a series of tiny biting motions. The song is an unmusical twitter. Best looked for at Abra Malaga.

Peruvian Sierra-Finch   *Phrygilus punensis*
15-16 cm. 2900-4600 m. A familiar bird of open country with scattered shrubbery on sloping or level ground, with rocky outcrops. Often encountered in towns, villages, farmyards, plowed fields and around houses. In pairs or small flocks outside the breeding season. Forages for seeds on the ground and perches in *Eucalyptus* trees and other bushes. Flight call is a sharp, often repeated *'tzit'* or *'tee-tit.'* The song consists of two alternating notes *'teer-zlip-teer-zlip.'* Common.

Mourning Sierra-Finch   *Phrygilus fruticeti*
17 cm. 2000-3800 m. Rather heavy bill and yellow legs. Inhabits bushy and rocky slopes in semi-arid areas, often with admixed cacti. Usually in pairs or loose groups, foraging on the ground and in bushes. Perches on rocks, exposed branches and the tops of the bushes. Male has a nuptial flight display, flying 2-3 m up before gliding down on vibrating wings whilst singing. Song is a dry wheezy *'wreeeee-iu-wreui.'* Fairly common at Huacarpay Lakes near Cusco.

Plumbeous Sierra-Finch   *Phrygilus unicolor*
14.5 cm. 3300-5000 m. Male all gray, female streaky. Inhabits rocky puna grassland and high Andean bogs with cushion plants, rocky outcrops and light *Polylepis* woodland. Usually encountered in pairs or small flocks feeding on the ground, and perched up on rocks and stone walls, often around houses. The song is a snarling *'wheezeee.'* Common at higher elevations.

Ash-breasted Sierra-Finch   *Phrygilus plebejus*
11-12 cm. 2400-4500 m. A nondescript sierra-finch and the smallest in the genus. Found in a variety of open habitats, specifically sparsely vegetated stony areas, bushy ravines and cactus covered slopes. Usually in pairs or groups. Often large flocks with other finches. Forages on the ground and perches on rocks and bushes. Common at Huacarpay Lakes and in the Cusco area.

Band-tailed Sierra-Finch   *Phrygilus alaudinus*
14-14.5 cm. 3000-4000 m. The *excelsus* subspecies is present. In our area local and uncommon in montane scrub and agricultural areas. Note in the male – white band in tail, yellow bill and contrasting white belly to rest of mainly grayish plumage. cf smaller Band-tailed Seedeater which has a smaller bill and rufous vent. The song is a high *'tszz-zzeww-zz-zweee-zzew.'* Can be seen at Huacarpay Lakes south of Cusco.

Short-tailed Finch   *Idiopsar brachyurus*
18.5-19.5 cm 3900-4600 m. All slaty gray with long large bill. cf Plumbeous Sierra-Finch which is shorter billed, smaller, more bluish gray and longer tailed. Locally found on talus slopes, boulder fields and the interface of *Polylepis* woodland with rocky slopes. Feeds on the ground among lichen covered rocks, often perches on boulders. Inexplicably, only known from the Yanahuara valley at Mantanay, which is less than 20 km from the well birded Abra Malaga area. Overlooked?

White-winged Diuca-Finch   *Diuca speculifera*
18-19 cm. 4000-5400 m. White crescent below the eye, white wing patch. Tail blackish with outer webs white (conspicuous in flight). Found on bogs with cushion plants at high altitude, as well as nearby rocky and grassy slopes. Has been recorded roosting in glacier cracks. Tame and found in pairs or family groups of 3-6 feeding on the ground in search of seeds. Flies long distances between feeding bouts and perches on rocks or stone walls to look around. Call is a snarling *'wheit.'* May be the highest nesting passerine in the Americas. Can be seen around the Salcantay massif and at Abra Malaga.

Slaty Finch   *Haplospiza rustica*
12 cm. < 3300 m. Male slate gray, female drab olive brown with streaks. Note long conical bill. Found in the understory of humid forest and forest edge, foraging for seeds on the ground in dense cover. Seems to be at least partially nomadic and very much tied to patches of seeding bamboo (*Chusquea* sp.) Male superficially similar to Plumbeous Sierra-Finch and Short-tailed Finch but is smaller and habitat totally different from those species. The song is a high pitched series of buzzy notes. Uncommon.

Grassland Yellow-Finch   *Sicalis luteola*
12-13 cm 3000-3700 m. Local. Male, drab and streaked, yellow superciliary and dark cheeks. Female similar but duller. Found in grassy fields and weed patches feeding on the ground. Song canary-like thin dry warbles and trills. Can be seen around Huacarpay Lakes. Probably under recorded.

Bright-rumped Yellow-Finch   *Sicalis uropygialis*
13-14 cm. 3330-4800 m. Gray cheek patch in both male and female. Found in open areas at high elevations in puna grassland, often with interspersed rocky slopes, also around small farms and llama corals with stone walls. Nests in stone walls, burrows or under the eaves of houses. Found in small to large flocks, often near water and loosely associated with sierra-finches and furnariids. Forages for seeds on the ground. The song is a descending series of twittering notes delivered from a rock or rooftop – *'de de de de dit it it it.'* Flight call is *'weet-weet.'* Can be seen at all high elevations in the area including Abra Malaga.

Greenish Yellow-Finch   *Sicalis olivascens*
13.5-14 cm. 2500-4200 m. Inhabits slopes with montane scrub and bushy ravines, often quite arid with cacti. Also, agricultural fields and pasture and around towns. Not in the puna zone and always lower than the preceding species and in a different habitat. Feeds on the ground in loose, often large, flocks. Its song is a loud series of 10-15 chirps, followed by a quite liquid *'tee-chee-chee che cheer te te.'* Flight calls include a clear *'tueet-tueet'* and a buzzy *'trruu.'* Easily seen around Cusco, Huacarpay Lakes and the Sacred Valley of the Incas.

Chestnut-breasted Mountain-Finch   *Poospiza caesar*
17 cm. 2500-3500 m. Peruvian endemic. A pretty bird of gardens, shrubby slopes and low woodland in semi-arid areas. Uses *Eucalyptus* trees and seems to prefer a mosaic of fields, bushes and trees. Skulking and can be difficult to see. Feeds on the ground underneath or not far from cover. Sometimes perches higher in *Eucalyptus* trees. The song is a chirping warble lasting 2-3 seconds and repeated every 10 seconds. Can be seen on the west side of Abra Malaga pass and near the Inca ruins of Tipon and Sacsayhuaman not far from Cusco.

Black and White Seedeater   *Sporophila luctuosa*
10.5-11 cm. < 2600 m. Inhabits mountain slopes in grassy areas with scattered trees and bushes interspersed with pasture. Also, along roads in the grass and hedgerows bordering agricultural areas as well as at the edge of humid forest. In pairs or small and nomadic large flocks, often with Yellow-bellied Seedeater. Forages for seeds in bushes and on the ground. The song, delivered from a high perch, is *'kieeer itititi teow-teow tow.'* Nomadic outside the breeding season.

Yellow-bellied Seedeater   *Sporophila nigricollis*
11 cm. < 2400 m. The *inconspicua* race is present at Machu Picchu. Many more female birds are seen than males. Inhabits shrubby and grassy clearings and the edge of humid forest, agricultural areas and roadsides. Encountered in pairs and flocks, often admixed with the previous species. The song is short and musical, often ending with a few more buzzy notes – *'tsee-tsee-tsee-bseeoo bzee-bzee'* with some variation. Can be seen around the Machu Picchu ruins.

Band-tailed Seedeater   *Catamenia analis*
11-12 cm. 2500-3700 m. Male: white wing speculum, rufous vent. In both sexes: tail blackish with white band across the center, more visible in flight. Found in agricultural areas, gardens, pastures and hedgerows, as well as scrubby, often arid, slopes, reeds and pastures. In pairs or small flocks, often with other finches. Forages for seeds in bushes or on the ground. The call is a dry *'stt.'* The song is a faint note followed by a buzzy *'tic bzzzz'* repeated at eight-second intervals. Common around Cusco.

Plain-colored Seedeater   *Catamenia inornata*
13-14 cm. 3000-4400 m. Male: bill pink to brownish pink; female: bill darker. Inhabits dry shrubby slopes, puna grasslands, isolated humid shrub with grassy areas and *Polylepis* woodland. Usually found in pairs, feeding mostly on the ground, sometimes in flocks outside the breeding season and sometimes with other finches. Voice is short buzzy trill. Uncommon at Abra Malaga.

Paramo Seedeater   *Catamenia homochroa*
12-13 cm. 2300-3800 m. Bill longer and more conical than in other seedeaters. Bright pale yellow bill looks shiny white at any distance. Found in undergrowth and on the ground in elfin forest and humid montane forest edge, and in grassy areas along roads. Usually encountered in pairs, sometimes with mixed species flocks of high elevation tanagers and flowerpiercers. Feeds on grass seeds. Flight call is *'tzit tzit.'* At Abra Malaga, but uncommon.

Bananaquit   *Coereba flaveola*
10-11 cm. < 2400 m. Short, sharply decurved bill. Inhabits forest edge, lighter woodland, second growth and gardens. An active bird, found alone, in pairs or in family groups, foraging at all levels, searching for nectar and fruit. The song is delivered from a high perch – a short, high-pitched series of hissing chips and buzzes. Mostly a lowland species but reaches 2400 m at Machu Picchu. Can be seen along the Urubamba River near Aguas Calientes and on the grounds of the Machu Picchu Pueblo Hotel.

Dull-colored Grassquit   *Tiaris obscura*
11 cm. < 2200 m. The bicolored bill, dusky above and yellowish below, is perhaps the best way to identify this nondescript grassquit. Can be overlooked as it resembles female *Sporophila* seedeaters. Song is a sudden *'zeetig zeezeezig'* with variations. Favors overgrown clearings, gardens, forest edge and verges. Usually in pairs or small groups not often associating with other seedeaters. At Machu Picchu usually encountered along the railway line or along the edge of the road leading from Aguas Calientes to the Machu Picchu ruins.

NEW WORLD SPARROWS   Emberizidae

Sparrows and brushfinches forage near or close to the ground mostly inside forest interior. Single birds and pairs forage in dense cover usually alone, but some *Atlapetes* will join mixed flocks. *Arremon* brushfinches usually hop along the ground whilst *Atlapetes* forage above the ground. *Chlorospingus* travel in pairs but mostly in small groups with or away from mixed species flocks in the mid and understory of humid forest.

Rufous-collared Sparrow   *Zonotrichia capensis*
14-15 cm. > 1800 m. A common and familiar Andean bird. Slightly crested. The immature is streaked and lacks any distinctive feature. Found in open and semi-open areas, natural or man-made, villages and towns. Tame and confiding, hopping on the ground, low walls, shrubs, etc. The song is often heard – a pleasant slurred whistle *'tee-teowww-tr-e-e-e-e'* often given after a rain-shower. The call is a sharp *'tic.'* Found at elevations up to 3600 m. Common even in city centers.

Yellow-browed Sparrow   *Ammodramus aurifrons*
13-14 cm. < 2000. Bend of wing bright yellow. Inhabits open areas, roadsides, agricultural areas, clearings. Quite tame, hopping about in open and low open shrubbery, feeding on the ground and low bushes. The song is high-pitched buzzy *'trick tzzee-zeee,'* repeated endlessly from an exposed perch. Mainly a lowland species and marginal at Machu Picchu. There are a few records from the Urubamba River between Mandor and the Aobamba Valley.

Chestnut-capped Brushfinch   *Arremon brunneinucha*
18.5-19 cm. < 2800. The *frontalis* race occurs at Machu Picchu. Note chestnut cap and contrasting bright white throat. Inhabits the undergrowth of humid montane and sub-montane forest, usually in the densest undergrowth, but sometimes moves out to the edge. Keeps low to the ground, often feeding on the ground, flicking aside leaves in the manner of a thrush. Usually not with mixed feeding flocks. The song is a very high-pitched series of notes followed by a trill. The contact note is a very high-pitched *'tzeet'* only audible at close range. Quite common along the Urubamba River near the Puente Ruinas railway station.

Gray-browed Brushfinch   *Arremon assimilis*
19-20 cm. 2000-3400 m. Gray superciliary with black breast band. Inhabits dense undergrowth in humid montane forest, forest edge and tangled undergrowth in mature secondary growth. Keeps close to the ground like the preceding species, which it generally replaces at higher elevations. The song is a high-pitched *'tzee-tzee-tse-weeee-tzee-szee-zee-trrrz,'* etc. Can be seen between Phuyuatamarca and Wiñay Wayna on the Inca Trail.

Tricolored Brushfinch   *Atlapetes tricolor*
16 cm. < 3000 m. Peruvian Endemic. Crown a rich yellowish gold. Sides of head black. Found in humid pre-montane forest, forest edge and well-developed secondary growth. Usually encountered in pairs or small family groups at low to mid levels. Generally unobtrusive, even in the company of mixed feeding flocks, but occasionally scans from the top of a low bush. The song is a series of slowly accelerating notes terminating with a dry trill *'tzit-tzit-tzit-tju-tju-tjud-trrrrrr.'* At the southern edge of its range at Machu Picchu. Fairly common in shrubbery around Machu Picchu ruins.

Apurimac Brushfinch   *Atlapetes forbesi*
18-19 cm. 2700-4000 m. Peruvian Endemic. Rufous head and nape with black around the eye. Fairly common where found in montane scrub and *Polylepis* woodland in intermontane valleys, in our area only likely at the base of the Salcantay massif. Song: low whistles and twitters, often delivered as a duet. One record from Peñas below Abra Malaga (vagrant?). Named for Sir Victor Walter Courtney Forbes (1889-1958) British diplomat in Peru.

Cuzco Brushfinch   *Atlapetes canigenis*
18 cm. 2450-3000 m. Peruvian endemic. The common brushfinch of the forests at Machu Picchu. A uniform dark gray bird with a chestnut crown and nape. Found in humid montane forest, adjacent forest edge and secondary growth. Usually in small family groups, often accompanying mixed species flocks, foraging at low to mid levels along branches and vines, sometimes in the canopy of small trees. When moving, maintains contact with other members of the group with a constant *'tzit-tzit'* note. The song is a series of chips and trills delivered from a high song-perch at dawn. Common between Phuyupatamarca and Intipunku along the Inca Trail and above San Luis at Abra Malaga.

Common Chlorospingus   *Chlorospingus ophthalmicus*
13-14 cm. < 2650 m. *Chlorospingus* were previously called bush-tanagers, but, in fact, are related to sparrows, not tanagers. The *peruvianus* race is present at Machu Picchu. Inhabits humid montane forest and borders, sometimes in lighter secondary growth. Forages in groups of 5-20, usually accompanying mixed species flocks. Searches leaves, branches and epiphytes for insects at all levels, also eats a lot of small berries. The calls are a variety of *'chip'* notes. The song is a distinctive series of stuttering *'chip'* notes, followed by a buzzy trill. Fairly common at Machu Picchu. Can be seen along the banks of the Urubamba River between Aguas Calientes and the Mandor Valley.

Yellow-whiskered (Short-billed) Chlorospingus   *Chlorospingus parvirostris*
14 cm. < 2750 m. Sides of the throat bright yellow, producing the effect of flaring yellow whiskers extending onto the ear coverts. Inhabits humid montane and sub-montane forest and forest edge. In pairs or small groups of 6-10, often with mixed species flocks also containing the previous species. Forages in the dense canopy and sub-canopy, hopping along mossy branches and perch-gleaning. Eats small fruits and insects. Constantly calls whilst foraging – *'tsip-seep.'* Can be seen along the Urubamba River in the lower elevations of the Machu Picchu Sanctuary.

CARDINAL GROSBEAKS Cardinalidae

Cardinal grosbeaks are robust, seed-eating birds with strong bills. They are an assemblage of distantly related songbirds. The word "grosbeak", first applied in the late 1670's, is a partial translation of the French *grosbec*, where *gros* means "large" and *bec* means "beak". The current sequence of species in this family is probably incorrect and will be re-evaluated at a future date.

Hepatic Tanager *Piranga flava*
17 cm. < 2700. Note dusky lores. The discontinuous range and ecological differences among subspecies have created uncertainty as to whether more than one species is involved. The *lutea* subspecies is present at Machu Picchu and may deserve full species rank as Highland Hepatic Tanager *(Piranga lutea)*. Bill darkish. Inhabits tall trees in humid montane forest, edge and lighter woodland, tall secondary growth alongside rivers, and sometimes montane scrub. Peers deliberately at foliage, twigs and underside of branches. Usually alone or in pairs, seldom with mixed species flocks. Call is an abrupt *'chup.'* The song consists of double notes – *'chp-whir weeti-chu wheet-cher,'* etc. Can be seen along the railway line between Puente Ruinas and the Mandor Valley.

Summer Tanager *Piranga rubra*
17-18 cm. < 3000 m. Boreal migrant present between the months of October to April. Very similar to previous species but lacks dusky lores. Heavy bill is pale. The immature male resembles the female but always shows patches of red in the plumage. Found in a variety of habitats, but generally in more humid situations than the preceding species. Usually encountered alone, rarely in pairs. Often inactive for long periods before dashing out and fly-catching for insects and berries. Often hover-gleans. The call is *'chika-tuk'* or *'chitik.'* Has been recorded along the Urubamba River between Aguas Calientes and the Mandor Valley.

Scarlet Tanager *Piranga olivacea*
17 cm. < 2200 m. Boreal migrant present Sept to March. Males arrive in dull basic plumage which is olive with yellow overtones and black wings, but acquires alternate red plumage with black wings before migrating north. Molting males are blotched red and green. This species has a smaller paler bill than the Summer Tanager. Canopy of forest and forest edge. Calls *'tchip'* and *'zweee'*. Rare. Has been recorded at Machu Picchu near Agua Calientes.

Golden Grosbeak *Pheucticus chrysogaster*
20-21 cm. < 3500 m. Formerly called Golden-bellied Grosbeak. Massive bill. Found in mostly semi-arid habitats, riparian thickets, shrubby slopes, gardens, *Eucalyptus* groves, hedgerows and fields. Usually alone or in pairs high in trees, though lower in some cases. Quietly forages for insects and berries, often in full view. Will raid corn fields. The call is a sharp *'kick.'* The song consists of gentle phrases alternating between two pitches *'hiuit hy huiu hye huii huit hy huit,'* etc. Rare.

Black-backed Grosbeak *Pheucticus aureoventris*
21-22 cm. < 3500 m. Massive bill. Inhabits humid montane forest and adjacent semi-arid zones in gardens, along streams with alders (*Alnus* sp.), and wooded ravines. Likes wild Andean cherries *(Prunus capuli)*. Alone or in pairs sometimes as part of mixed feeding flocks. Mainly forages for berries and insects at mid-levels. The song is a series of melodious and varied whistles. Call is a sharp *'kick.'* Fairly common at Machu Picchu and in the Sacred Valley of the Incas.

·TAWA RESTAURANTE, YUCAY, PERU
 SACRED VALLEY
 12/27

NEW WORLD WARBLERS   Parulidae

These small, active, insectivorous birds can be divided into two groups – migrants from North America that pass the northern winter in the south, and resident species dominated by the difficult *Myiothlypis* genus. Warblers inhabit forest and are often a feature of mixed feeding flocks. The song is given by the male. Nest building and incubation is by the female only, but both parents feed the young. The nest is a cup placed in a shrub or bush, or sometimes a domed nest in a cavity or on the ground.

Masked Yellowthroat   *Geothlypis aequinoctialis*
13-14 cm. < 2000 m. Male distinctive but female confusing with few field marks. Inhabits dense thickets and bushy vegetation at the edges of clearings, rivers or marshy spots. Keeps low to the ground, moving constantly, hopping up and down between thin branches, circling and cocking its tail. Difficult to see unless the male is singing from an exposed perch – a pleasant warbling song, rather finch-like in character. Rarely seen at Machu Picchu where it may be a rare austral migrant.

Cerulean Warbler   *Setophaga cerulea*
12 cm. < 2000 m. Boreal migrant present September to March in Peru. Black necklace and streaked sides of the male are distinctive. Female and immature always show unstreaked blue-green back and white superciliary. Forages in mid to upper level of humid forest often with mixed species flocks. Call a buzzy *'zit.'* One record from near Agua Calientes but more could be expected at the right time of year.

Tropical Parula   *Setophaga pitiayumi*
10.5-11 cm. < 2200 m. Dull blue above, yellow below with an olive patch on the mid-back. Vent white. Shows two prominent white wing bars. At Machu Picchu inhabits humid montane forest and borders, foraging at mid to high levels, often with mixed flocks of tanagers. The often-heard song is a thin, buzzy trill – *'sip-sip-sip-sip-sipstriiip,'* accelerating toward the end. Common around Aguas Calientes, particularly on the grounds of the El Pueblo Hotel.

Blackburnian Warbler   *Setophaga fusca*
12-12.5 cm. < 3000 m. Boreal migrant present from September to early April. Mostly, but not always, seen in non-breeding plumage at Machu Picchu which is similar to the female's plumage. Found in the forest canopy and borders, mature second growth woodland and clearings with scattered trees. Mostly alone or in pairs, but sometimes in larger loose groups with mixed feeding flocks of tanagers etc. Feeds at mid to high levels. Can be seen along the Urubamba River between Aguas Calientes and the Mandor Valley.

Citrine Warbler   *Myiothlypis luteoviridis*
13.5-14 cm. 2500-3700 m. The *stratiaceps* race is present at Machu Picchu. Similar to the following species which it generally replaces at higher elevations. Legs yellowish-brown. Very similar to Pale-legged Warbler, but note the longer yellow supercilium and blackish sides of the crown. Found in lower growth of montane forest and elfin forest, borders and second growth. Usually in pairs or small family groups, low to the ground, foraging by perch gleaning or short sallies. Often found in mixed feeding flocks. The song is similar to Pale-legged Warbler, perhaps more musical. Can be seen easily on the Inca Trail between Sayacmarca and Phuyupatamarca and near tree line at Abra Malaga.

Pale-legged Warbler   *Myiothlypis signata*
13-13.5 cm. < 2800 m. The nominate *signatus* race is present at Machu Picchu. Dark line through the eye and a short yellow supercilium. Legs noticeably pale. Inhabits lower growth of sub-montane forest, woodland and forest borders, often with admixed bamboo. Keeps close to the ground in pairs or small family groups, often with mixed understory flocks. Perch gleans, picking insects off the surface and, particularly, the underside of leaves. Sally gleans also. The song is a distinctive series of fast *'chippers,'* rising then dropping in pitch as volume increases. Common at Machu Picchu, especially near Puente Ruinas railway station and the grounds of the El Pueblo Hotel in Aguas Calientes.

Russet-crowned Warbler *Myiothlypis coronata*
13.4-14 cm. < 2900 m. Separated from the preceding two species by an orange-rufous crown, bordered by a black stripe, continuing onto the nape and the lack of yellow in the throat. Found at low to mid-levels in the understory of humid montane and sub-montane forest as well as mature secondary growth. Usually quite easy to see, as they forage in circles along thick horizontal branches. Sometimes in mixed species flocks but just as likely to be seen in pairs. The pretty song is a musical and pleasant warble rising at the end *'tlee-tlee-tlee-telee-wheedeee-weeii.'* Quite common along the Urubamba River between Aguas Calientes and the Aobamba Valley.

Three-striped Warbler *Basileuterus tristriatus*
12-13 cm. < 2500 m. Long supercilium and black cheeks. Inhabits the understory and edge of humid montane forest, generally foraging between 1 m and 3 m, sometimes wandering into bushy clearings and secondary growth. In pairs or small groups, often accompanying mixed flocks. Searches thickish branches twisting its body from side to side. The song is a sputtering, twittering variety of notes, not very distinctive.

Canada Warbler *Cardellina canadensis*
Length 12-13.5 cm. < 2500 m. Boreal migrant present from September to early April. Males mostly seen in non-breeding plumage which is similar that of the female. In dull females and immatures the necklace may not be visible. Found in the lower growth of forest and secondary woodland. Active and restless, gleaning insects from foliage. Often accompanies mixed feeding flocks. Uncommon at Machu Picchu and best looked for along the Urubamba River between Aguas Calientes and the Mandor Valley.

Slate-throated Redstart (Whitestart) *Myioborus miniatus*
12-13 cm. < 2600 m. The name redstart is derived from the superficial overall similarity to Eurasian *Phoenicurus* redstarts (which are also not red). Some authors prefer 'whitestart' but there are convincing arguments for retaining redstart. Redstarts are long-tailed, confiding and conspicuous members of the warbler family. Shows a small semi-concealed chestnut crown patch. Tail blackish, with outer tail feathers white, this latter feature very noticeable. Found in humid montane forest and woodland, borders and mature second growth. Feeds mostly at mid to high levels but descends on occasion. Usually in pairs or small groups, they are a conspicuous element of mixed feeding flocks. Actively perch gleans, wagging its body from side to side and flashing white outer tail feathers. Also fly-catches. The often heard song is an accelerating series of *'tswit-tsweet'* notes. Seems to be nuclear species of montane mixed flocks at elevations mostly below 2500 m. Common at Machu Picchu and easily seen along the railroad track between Aguas Calientes and the Mandor Valley.

Spectacled Redstart (Whitestart) *Myioborus melanocephalus*
13-13.5 cm. 2000-3500 m. The *bolivianus* race is present at Machu Picchu. Crown and sides of head black with bright yellow forehead and ocular area (spectacles). Generally replaces the preceding species at higher elevations but much overlap with both species occurs in the same mixed flocks at certain elevations at Machu Picchu. Found in humid montane forest and shrubbery, sometimes in elfin forest. Like the preceding species, it seems to be a nuclear species in mixed feeding flocks. Found in groups of 4-5 or in pairs. Perch gleans, twisting body from side to side and often fanning tail, flashing white outer tail feathers. Also fly-catches. The often heard song is an unremarkable twitter of notes. Common near Wiñay-Wayna ruins along the Inca Trail and at Abra Malaga.

ORIOLES AND BLACKBIRDS   Icteridae

A varied group confined to the New World reaching their greatest diversity in the tropics. Most are medium-sized to large birds, with the females being usually smaller and/or duller than the males. Bills are typically conical and sharply-pointed, and they have powerful feet. Many are noisy, gregarious and conspicuous birds. Some species, such as the oropendolas, construct intricately-woven, long, pendulous hanging nests in colonies. They eat mainly fruit and seeds. Display and courtship rituals are usually elaborate.

Dusky-green Oropendola   *Psarocolius atrovirens*
41 cm. (Male). 33 cm. (Female). < 2600 m. Bill pale greenish-yellow. Central and outer two pairs of tail feathers olive, the rest yellow narrowly-tipped with olive. Inhabits the borders of humid sub-montane forest and adjacent clearings with scattered trees and along streams and rivers. Nests in small colonies and pairs typically forage for fruit and berries in the canopy. Males displaying at a colony "fall" forward on their perch with flapping wings and raised tail. Calls include a loud liquid *'chook'* and *'jog.'* Conspicuous and common along the Urubamba River between Aguas Calientes and the Mandor Valley.

Southern Mountain Cacique   *Cacicus chrysonotus*
Length 30-31 cm. (Male). 25-26 cm. (Female). < 3450 m. Bill long and pale gray. Iris pale blue. Mostly black with lower back and rump yellow (only visible in flight). Inhabits humid montane and sub-montane forest and forest edge, often where there are large stands of bamboo *(Chusquea)*. In pairs or flocks of up to 10 birds, sometimes joining other large birds such as Hooded Mountain-Tanagers and White-collared Jays. Climbs about in the canopy of the trees searching for insects, berries and nectar from flowering trees. Often lower in bushes and bamboo where it investigates internodes. Flies long distances between feeding localities. Quite vocal. Calls include a clear whistle *'peew'* and a quavering *'wee-eeuuuuuw.'* Many other varied calls. On contact it emits a harsh *'chack.'* Can be seen near Wiñay Wayna ruins along the Inca Trail and at Abra Malaga.

Yellow-billed Cacique   *Amblycercus holosericeus*
21.5 cm. (Male). 21 cm. (Female). < 3200 m. Uniform dull black. Bill pale greenish yellow. Found in dense undergrowth at the edge of humid montane and sub-montane forest, often where there are large patches of bamboo *(Chusquea)*. Usually encountered in pairs or small family groups, often with mixed species flocks. Scratches, tears and rummages among trapped dead leaves, mosses, tangles and dry stems. Very unobtrusive, staying well hidden and usually only seen as it ascends undergrowth into flowering trees after nectar or flies across a trail or clearing. Its call is a harsh *'waak.'* The song is a series of piercing whistles.

Yellow-winged Blackbird   *Agelasticus thilius*
18-18.5 cm. 3000-4300 m. Male black with concealed yellow shoulder patch which becomes conspicuous in flight. Female and immature gray-brown and heavily streaked. Common in reed beds around lakes near Cusco and adjacent pastures and tall crops. Song a varied variety of wheezy note, trills and whitles. Call *'chink'*. Common at Huacarpay and Huaypo Lakes near Cusco.

Bobolink   *Dolichonyx oryzivorus*
17-18 cm. < 2500 m. Boreal migrant. A highly migratory *Icterid*, only transient in our area: October to November and March to April. Breeding male is black below and mostly white above, a tuxedo on backwards as Roger Peterson would say. Male breeding plumage appears gradually, and is more evident during northward passage (March to May). A bird of grasslands, during migration, usually noted in flight making its characteristic *'ink'* call note. Will drop into clearings and along rivers on migration. Its migratory patterns are little understood and the bird appears to be overlooked over most of its migratory route.

White-browed Meadowlark (Blackbird)   *Sturnella superciliaris*
18 cm. 3000 m. Rare austral migrant to lowlands with one record from Huacarpay Lakes near Cusco. Male black and red with prominent white superciliary. Female probably not separable from Red-breasted Blackbird which could also occur in the area as a wanderer. Found in open habitats, fields and agricultural areas.

FINCHES   Fringillidae

Euphonias live in pairs and are unique among birds in lacking a gizzard. They regurgitate food for nestlings. They have melodic calls and build a round nest with a side entrance hole. Siskins are small finches, with males boldly-patterned in black and yellow. Females duller and more olive. Quite difficult to differentiate between species. Gregarious and mostly found in open or semi-open terrain, usually feeding in trees and shrubbery, sometimes on the ground and sometimes in very large flocks.

Thick-billed Siskin   *Sporaga crassirostris*
13-14 cm. 3600-4600 m. A large siskin resembling the Hooded Siskin in plumage. Differs in having a noticeably thicker bill. Center of belly white. Female grayer than female Hooded Siskin without any yellow except in the wing. Inhabits *Polylepis* woodland and nearby bushy slopes and ravines. Found in pairs or small groups. Feeds on seeds and buds of *Polylepis*. Only likely to be found where stands of *Polylepis* exist such as at Abra Malaga.

Hooded Siskin   *Sporaga magellanica*
11-12 cm. < 4200 m. The most familiar and widespread siskin. Wings black, with a yellow band on the greater coverts and also across flight feathers (conspicuous in flight). Found in humid and semi-arid shrubbery, gardens and agricultural areas with hedgerows, in small to large groups foraging at all levels for seeds. The song is a continuous twitter. Tends to avoid dense humid forest where it is replaced by the next species. Quite common in the agricultural fields in the Sacred Valley of the Incas.

Olivaceous Siskin   *Sporaga olivacea*
11 cm. < 2500 m. Extremely similar to Hooded Siskin, being more strongly tinged olive in both sexes and appearing slightly smaller. Perhaps best identified by habitat. Found in humid montane forest clearings and edge – this habitat is usually avoided by the Hooded Siskin. It appears that where both species co-exist, as at Machu Picchu, they maintain distinct habitat preferences. The song is similar to that of the Hooded Siskin. Can be seen along the Urubamba River between Aguas Calientes and the Mandor Valley.

Black Siskin   *Sporaga atrata*
12-13 cm. 3500-4800 m. Female like the male, but dull black as opposed to glossy black. Found in the puna zone on rocky slopes and in ravines with small bushes, along stone walls and near human habitation. Usually encountered in small groups feeding on the ground. Song typical of the genus, delivered in nuptial song flight or from the top of a bush with drooping wings. Can be seen at Abra Malaga.

Thick-billed Euphonia   *Euphonia laniirostris*
10 cm. < 2200 m. The *melanura* race is present at Machu Picchu. Forehead of male is yellow to just behind the eye. Underparts yellow up to the bill. Inhabits forest borders and secondary growth with some tall trees. Usually encountered in pairs often with mixed feeding flocks of other tanagers, foraging at all levels but mostly mid-levels. Often rests on exposed dead branches. Searches for berries. Calls are loud and musical, with a downward inflection – *'chwee'* or *'tweer'* or *'wheet'* or *'chee-weet.'* Often imitates other birds with great accuracy. The song is soft, musical and varied. Can be seen near Puente Ruinas railway station and the ground of the Machu Picchu Pueblo Hotel.

Orange-bellied Euphonia   *Euphonia xanthogaster*
11 cm. < 2200 m. The *brunneifrons* race is present at Machu Picchu. Shows white under tail spots on a black tail. Female: mostly olive above with dull rufous forehead. Buffy-gray below, buffier on the belly. Sides and flanks yellowish-olive. Inhabits humid sub-montane forest, forest borders and edge and well-developed secondary woodland. Usually alone or in pairs, occasionally with mixed species flocks and often with feeding aggregations at fruiting trees. Forages at all levels but not too high, and often quite low in the forest interior. Frequent calls include *'deet-de-deet'* or *'chhee-wheet'* and variations on this theme. Can be seen along the Urubamba River between Aguas Calientes and the Mandor Valley.

Blue-naped Chlorophonia   *Chlorophonia cyanea*
11 cm. < 2000 m. The nominate race is present at Machu Picchu. Note green head, blue collar and yellow underparts. Juveniles are green, tinged yellow below. Inhabits humid sub-montane forest, mostly at the forest edge and in clearings, in family groups or pairs, often associated with mixed feeding flocks. Often joins feeding aggregations at fruiting trees. Actively and deliberately picks fruit and insects from trees, usually in the canopy. Calls include a nasal *'ek'* or *'enk,'* also a downward slurred *'teeeu.'* Also a short rattle. All these notes are sometimes mixed together. Best looked for at the lowest elevations within the Machu Picchu Sanctuary.

Thick-billed Euphonia

Orange-bellied Euphonia

Blue-naped Chlorophonia

A BIRD FINDING GUIDE
TO THE CUSCO AND MACHU PICCHU AREA
AND
AN ANNOTATED CHECKLIST

A BIRD FINDING GUIDE TO THE CUSCO AND MACHU PICCHU AREA

Birds are everywhere. If this is your first trip to the Andes, you will have ample opportunity to see new species, even if birdwatching is not the principal focus of your trip. If birds are your main reason for being here, do not ignore the impressive Inca archaeological sites and rich Andean culture found in the Cusco area and at Machu Picchu. This guide will help you plan half day or full day escapes to see some of the special and widespread birds of the area. Of course, if you hire a bird tour leader in Cusco or take a trip with a travel agency specializing in birding and wildlife tours, your chances of seeing most of the endemics and a good number of species will increase greatly. Generally, such arrangements need to be made in advance. However, if you want to search for birds on your own, most of these sites are accessible by hiring a taxi or booking a reliable car and driver through a travel agency or hotel reception. Rental cars are available in the city, but having a driver to look after the car and your gear is a good idea to insure worry free days birding. At Machu Picchu, sites may be reached by walking from your hotel. Endemics are indicated thus: (E).

WEATHER

The Highlands of Cusco and Machu Picchu have two distinct seasons. The rainy season is during austral summer – October to April – and the dry season coincides with austral winter – May to September. The most rain falls in January and February, however, with global climate shift it is getting harder to predict the seasons. The rainy season is not a monsoon and some days can be rainless, others rain free in the morning with cloud build up and rain in the afternoon. Daily temperatures in Cusco are on average moderate with few extremes – average highs are from 10-13° C and nightly lows 1- 7° C. June is the coldest month. Temperatures rise or drop depending on the elevation. As a rule of thumb, let's say you are in Cusco Plaza at 3400 m and it is 13° C. If you were at Abra Malaga pass at 4340 m it would be 9° C and at Machu Picchu Ruins at 2425 m, 20° C. During May-July at dawn it is well below freezing at Abra Malaga Pass. This mini bird-finding guide covers an elevation range of 2000 m to 4300 m, so varying climates will be encountered. Sunrise and sunset vary little from midsummer (December) when days are longer (5:10 am to 6:10 pm) to midwinter (June) when the days are shorter (6:10 am to 5:30 pm). Layers are the way to go: chilly early in the day and warming up as the day progresses.

Approach to Huacarpay lakes heading southeast from Cusco

SOME BIRDING SITES IN THE AREA

Cusco and the surrounding area

Cusco (sometimes spelled Cuzco) is the gateway to Machu Picchu and a major tourism centre. Almost everyone who goes to Machu Picchu must pass through the city which boasts a historic town centre with intact Inca architecture and many hotels and restaurants for all budgets. You can see many birds casually while exploring local Inca sites such as Sacsayhuaman, Tambomachay and Quenko though we do not specifically mention them here.

Huacarpay Lakes

Perhaps the most well-known birding location close to the city is just 30 km south of Cusco on the main road to Puno and Bolivia. A half a day (with car) is enough to cover this area though a full day at a leisurely pace will get you a few more species. Many species are resident all year round and the wetlands are a magnet for migrating boreal migrants, particularly from September to October and April to May.

ACCESS AND STRATEGY: Leaving Cusco on Avenida de la Cultura to the southeast drive, take a taxi or colectivo (shared vans or taxis) to Huacarpay. Colectivos depart from Avenida de la Cultura opposite the University/Hospital Regional area. After 30 km, you will see the lakes on your right. Huacarpay Lakes are situated next to the town of the same name just before the turn off to Paucartambo and the world famous Manu Road. The lakes are surrounded by Inca and Wari archaeological sites.

A road circles the main lake – Waton – and starts at 2970 m [S13 36.618 W071 44.157] – just past the Sagitario service station – and this is where you should bird, paying attention to the reed beds and open water. The full circuit of the lakes is a long walk if you are on foot – just less than 8 km. Most all of the passerine birds will be found on the scrubby xerophytic hill slopes surrounding the lake – a good area is near the restaurant at Urpicancha (open weekends, basic accommodation) at [GPS S13 37.462 W071 43.109] 3100 meters. Just walk up the hillside where you can, a few hundred meters either side of the restaurant. Alternatively, the nearby ruins associated with the archaeological site of Pilkillacta on the east side of the main road have good areas of arid montane scrub and another smaller wetland known as Choqepuquio [GPS S13 36.194 W071 43.755] – watch out for the dogs and ask the local landowner for permission. A word of warning: The lakes are a favorite picnic spot on weekends and can

Approach to Tipon exit heading southeast from Cusco

be crowded and busy then, the upside being several local and very traditional restaurants are open for lunch in the nearby colonial town of Lucre – Duck a specialty (farmed)!

Key species: Bearded Mountaineer (E), Rusty-fronted Canastero (E), White-tufted Grebe, Yellow-billed Teal, Yellow-billed Pintail, Cinnamon Teal, Puna Teal, Andean (Ruddy) Duck, Puna Ibis, Little Blue Heron, Black-crowned Night-heron, Cinereous Harrier, Black-chested Buzzard Eagle, Variable Hawk, Mountain Caracara, Plumbeous Rail (reeds), Andean Lapwing (mostly winter) Andean Gull, Andean Coot, Bare-faced Ground-Dove, Black-winged Ground-Dove, Andean Swift, Sparking Violetear, Giant Hummingbird, Black tailed and Green-tailed Trainbearer, Andean Flicker, Wren-like Rushbird (reeds), Streak-fronted Thornbird, White-crested Elaenia (mostly winter), Yellow-billed Tit-Tyrant, Many-colored Rush-Tyrant (Reeds), White-browed Chat-tyrant, Spot-billed and Rufous-naped Ground Tyrants (winter), Andean Negrito (lake edge), Chiguanco Thrush, Brown-bellied Swallow, Hooded Siskin, Cinereous Conebill, Peruvian, Mourning and Ash-breasted Sierra-Finches, Greenish and Grassland Yellow-Finches, Band-tailed Seedeater, Black-throated Flowerpiercer, Golden-billed Saltator, Yellow-winged Blackbird (reeds) and, depending on the time of year, a variety of boreal shorebirds including Buff-breasted Sandpiper and Upland Plover. The wetlands attract many vagrants and almost anything that migrates may turn up.

Tipon ruins

On the way to or from Huacarpay Lakes this short detour can get you another Peruvian endemic – the pretty Chestnut-breasted Mountain-Finch.

ACCESS AND STRATEGY: Driving south from Cusco you pass through Saylla (14 km – a good place for wok fried pork and crackling!). Continuing to kilometer 22, you arrive at Tipon and the turn off to the town and ruins [GPS S13 35.26 W071 47.86]. This is about 13 km before Huacarpay Lakes. Take this road and continue through town bearing left at the pretty church and plaza, to the entrance gate to Tipon Ruins [GPS S13 34.307 W071 47.058] 3400 m - these are certainly worth a look and birds can be seen around this archaeological site. However, if you bird the first few bends below the car park at the entrance gate, you may see all species mentioned. The Peruvian Pygmy-Owl can often be enticed out of the stands of *Eucalyptus* trees. [GPS S13 34.471 W071 47.058] 3330 m is a good stop.

Key species: Spot-winged Pigeon, Peruvian Pygmy-Owl, Bearded Mountaineer (E), Chestnut-breasted Mountain-Finch (E) and other passerines mentioned for Huacarpay.

Approach to right turn to lake Piuray heading north

View of southeast corner of lake Piuray

Lakes Piuray and Huaypo

Both these lakes attract good numbers of water birds and are located close to the road from Cusco to Chinchero and the Sacred Valley of the Incas. As they are at a higher elevation than the Huacarpay Lakes, they hold some species not found at that site. They are an interesting stop on the way to the Sacred Valley of the Incas and worth an hour or two. Telescope useful here.

ACCESS AND STRATEGY: Drive northeast out of Cusco over Tika-Tika pass for 49 km toward Chinchero. You soon arrive at Poroy, beyond which the road diverges at the Petromundo service station – left to Nazca and Lima and right to Chinchero and Urubamba. Go right and immediately before Chinchero you will see some small marshy ponds on the left and right which are worth a look and contain Many-colored Rush Tyrant, Wren-like Rushbird and Plumbeous Rail.

LAKE PIURAY 3670 m: There are two access roads to the lake – the first you come to is at [GPSS13 25.209 W072 03.516] and is somewhat hidden between houses- this will take you directly to the southeast corner of the Lake after 1.5 km on a rough track. The second is right at the entrance to Chinchero itself (you will notice Piuray Lake on your right as you approach Chinchero). Turn right at [GPSS13 24.384 W07203.201], slightly doubling back on yourself and then right again – this road will take you along the west shore and eventually to an old pumping station in the SW corner. A road circles the lake and the lakeshore and water are visible at various points, but perhaps the best birding areas are on the west and south-west shores of the lake. There are some interesting Inca ruins at Cuper Bajo on the east shore of the lake.

LAKE HUAYPO 3500 m: After returning to the main road, continue for 1 km through Chinchero (colorful Sunday market and Inca ruins on the main plaza) and continue heading toward Urubamba and the Sacred Valley of the Incas. A new and unnecessary airport for Cusco is planned for this stretch which, if built, will destroy one of the most beautiful landscapes in the Peruvian Andes. After 4.5 km and a winding stretch of road, you arrive at a sharp left-hand hairpin turn with the lookout point of Raqchi and car park on the right at [GPS S13 21.848 W072 04.076] 3770 m. (Definitely worth a stop for a photo). Continue for 7 km more and turn left (the lake is signposted) at the village of Cruzpata at

Approach from Chincero to left turn to lake Huaypo

[GPS S13 22.594 W072 07.162] 3600 m. Continue straight, avoiding turns, to the lake shore which will be on your left. Perhaps the best area is the north and northeast corner. Make your way to the lake shore by picking your way through the corn and bean fields or continue farther along to where the road is closer to the lakeshore and bird from the vehicle.

Key species: Silvery Grebe (mostly winter) White-tufted Grebe, Andean Goose (mostly winter), Crested Duck (mostly winter) Yellow-billed Teal, Yellow-billed Pintail, Cinnamon Teal, Puna Teal, Andean (Ruddy) Duck, Neotropic Cormorant, Puna Ibis, Andean Ibis, Cinereous Harrier, Black-chested Buzzard Eagle, Variable Hawk, Mountain Caracara, Plumbeous Rail (reeds), Andean Lapwing, Puna Plover (rare), Collared Plover (rare), Andean Gull, Andean Coot, Eared Dove, Spot-winged Pigeon, Andean Flicker, Wren-like Rushbird (reeds), Many-colored Rush-Tyrant (reeds), Spot-billed, Rufous-naped and other Ground Tyrants (winter), Andean Negrito (lake edge), Chiguanco Thrush, Brown-bellied Swallow, Short-billed and Correndera Pipits, Hooded Siskin, Cinereous Conebill, Peruvian and Ash-breasted Sierra-Finches, Greenish Yellow-Finch, Black-throated Flowerpiercer, Golden-billed Saltator, Yellow-winged Blackbird (reeds) and, depending on the time of year, a variety of boreal shorebirds. The wetlands attract many vagrants and almost anything that migrates can turn up.

The Sacred Valley of the Incas

What today is popularly known as the Sacred Valley of the Incas is in fact the valley of the Urubamba River (Wilka Mayu to the Incas), the river of which continues downstream past Machu Picchu. It is generally understood to include everything between Pisac (33 km from Cusco) at 2846 m to Ollantaytambo at 2800 m elevation, about 55 road kilometers to the northwest. A passage from one to the other would take you through Lamay, Calca, Urubamba and Yanahuara. There are many hotels and restaurants mainly clustered in Pisac, Urubamba, Yanahuara and Ollantaytambo. Although the heavily cultivated valley floor and the river itself are of little interest to birders, the Urubamba is fed by numerous streams which descend through adjoining valleys and gorges of the Vilcanota massif, where numerous archaeological remains and villages are found. The valley was appreciated by the Incas due to its special geographical and climatic qualities and was the heartland of the Inca Empire. Here I will mention some, but not all, of the accessible side valleys entering from the Cordillera Vilcanota to the northeast. Not frequently visited by

View of lake Huaypo from the north

birders, these sites hold some interesting species and contain tracts of endangered Queñua or *Polylepis* woodland. Ideally, you will need your own vehicle for these valleys.

ACCESS AND STRATEGY: The Sacred Valley of the Incas is accessed by two roads from Cusco. One winds 33 km up past the megalithic Inca fortress of Sacsayhuaman and the Quenko and Tambomachay ruins, over the Corao pass and down to Pisac in the valley. The second leaves Cusco over the TikaTika pass and goes to Chinchero, passing close to Lakes Piuray and Huaypo, and onto Urubamba in the valley floor – around 57 km.

PISAC TO COLQUEPATA (33 km from Cusco): From the Pisac bridge, take the road that winds up to the Pisac Ruins (you will pass the Royal Inca Hotel on your left after 1.4 km) continuing on past giant Inca terracing on your left, do not continue to the ruins (unless you want to see them – impressive) but turn right after 6.5 km on a dirt road at [GPS S13 20.371 W071 47.146] to Amaru and Quello Quello. At 12 km there is a road split – right to Cuyo Grande and on to Colquepata, left to Pampallacta. These roads continue to high puna grasslands grazed by alpacas and llamas, passing through mosaics of cultivated fields and hedgerows. Colquepata is 35 km from Pisac and you may continue to Paucartambo, where you can join the road to the Manu Biosphere Reserve. There is not much habitat low in the valley but higher, around small farms and up to the puna grasslands, there is always something to see. It is worth birding to the highest points and back again. It is only worth going beyond Colquepata if you intend to continue to Paucartambo and Manu. Both are good dirt roads and a good immersion into traditional Andean culture.

THE LAMAY VALLEY (11 km from Pisac): From the village of Lamay at 2970 m, a reasonably good dirt road ascends through different elevational zones to high puna grassland before descending on a steadily worsening road to Challabamba on the road to Manu. Take the road that borders the right-hand side of the pretty plaza which turns off the main Pisac-Urubamba Road at [GPS S13 21.933 W071 55.296]. It is worth birding to the highest point and back again, but only worth going beyond the pass at [GPS S10 17.419 W071.49.98] 4338 m if you intend to continue to Manu (4 x 4 recommended). From the main road to the pass is 22.6 km. The road ascends via the bridge at Sayhua (3 km), passes the signed turn off to Huamay (7 km) and onto the village of Poques (12.6 km). Below Poques you are in a mosaic of agricultural farmsteads, scrubby arid hillsides and *Eucalyptus* groves. Above Poques you are onto high puna grasslands grazed by llamas and alpacas with scattered lakes and bogs. Between Poques and

Approach to right turn to Lamay valley northwest of Pisac

Sapaccto at kilometer 16.5 [GPS S10.17.725 W071 52.103 3960 m] an obscure rough track leads off to the left to a beautiful high elevation lake named Pacchar which, as well as hosting other waterfowl, has breeding Giant Coots. It is not far and you might want to leave your car at the start of the track and walk (just over 1 km as the caracara flies). Continuing to Sapaccato, the grasslands around the village, corrals and gardens are attractive to Andean Flickers, Sierra-Finches, Yellow-Finches, Cinclodes, Canasteros, Common and Slender-billed Miners and, especially in winter, a variety of Ground-Tyrants.

THE CALCA VALLEY TO AMPARAES PASS (7 km from Lamay): This is a paved road that leaves Calca at 3034 m. Turn off on the main road signposted to Lares and Baños Termales de Machacancha taking

Approach to turnoff in calca to Calca valley and Ampares Pass

Approach to right turn in Urubamba to Urubamba valley

the road that winds its way up for 30 km to the Amaparaes pass at [GPS S13 10.803 W071 54.535] 4514 m. After 8 km you reach the hot springs of Machacancha and after 19.5 km the extensive ruins of Ankasmarca on your left with over 150 Inca storage buildings (worth a visit). After 23.3 km at [GPS S13 12.454 W071 54.704] there is a turn off to Lares on your left, the road to the right goes to Amparaes pass and continues to Quebrada Honda and eventually, by a back route, to Quillabamba in the lowlands. It is a good half day trip to the pass and you can stop anywhere you see vegetation and grasslands. The bird community is similar to other valleys in this section.

THE URUBAMBA VALLEY (19 km from Calca): Driving along the main road through Urubamba turn right at the big gas station in town at [GPS S13 18.566 W072 06.869] at 2830 m. Continue up the main street past the cemetery on your right. Keep on going as far as you can, birding once you are outside of town. The road dead ends below the Chicon Glacier and you must return the same way. Alternatively, from behind the Church of the Señor de Torrechayoc [GPS S13 18.094 W072 07.591] 2946 m drive keeping right along the road to the Pumahuanca valley where some semi-humid scrub can be productive. Again this road dead ends just past the Chupani Inca Ruins at [GPS S10 15.679 W072.07.591] 3253 m. You must return the way you came. The bird community here is similar to other valleys in this section, but it is a good place to look for Crested Becard.

THE YANAHUARA VALLEY (8 km from Urubamba): Turn right at Yanahuara village and follow signs to one of the two hotels mentioned below (one turn off is at [GPS S13.16.485 W072 10.404] 2892 m) and continue up the valley. There are two nice hotels in this valley on the outskirts of the village of Yanahuara both with birdy gardens, the Casa Andina Private Collection and the Hacienda del Valle Hotel. Just past this latter hotel the road ends at [GPS S13 18.092 W072 10.462] 3182 m, but a wide mule trail continues to the high Puna and the unpronounceable Huacahausijasa Pass. This is for the adventurous only. Below the pass at Mantanayat around [GPS S13 14.360 W072 16.911], there are fairly large tracts of *Polylepis* woodland which hold all the Queñua *or Polylepis* birds mentioned for the Abra Malaga *Polylepis*, but here there are perhaps more territories of each species. It is a tough 4 hour hike to the woodland but you might be able to arrange horses and make it in an hour and a half. There are many patches of woodland and the higher ones are the best for Royal Cinclodes. An isolated population of Short-tailed Finch exists here. This is one of the few valleys in the area that passes through intact woodland types from 3000 m to tree line. Unless you have horses, overnight camping might be in order.

View of turnoff to Yanahuara valley in Yanahuara village

THE PATACANCHA VALLEY TO OCCOBAMBA (12 km from Yanahuara): The same hotels mentioned in the Abra Malaga section can be used as a base for birding this valley. Passing the main plaza in Ollantaytambo, make a tight right turn just before reaching the bridge over the Patacancha River. Once on this road there are no major turn offs and the road ascends past megalithic Inca terraces and other Inca sites. After 9.4 km you reach Markacocha [GPS S13 13.335 W072 12.395] 3316 m and its charming church, then the village of Huilloc [GPS S13 12.260 W072 12.052] 3649 m and 4.6 km farther on Patacancha, each village with its own ancient traditional weavings. To reach the high pass of Yuricunca at 4451 m it is an additional 14.5 km [GPS S13 06.695 W072 13.628]. And from here you can continue on over the high pass to Occabamba in the cloud forest (2 hours or so). There is a new road we have not explored branching off to the right 8.5 km before the pass at [GPS S13 08.131 W072 11.852] 4137 m that connects to Lares. Stop anywhere you see habitat and birds right up to the pass. Birding to the pass and back can be done in a day but beyond that you would need to be self sufficient and camp. The birding on the way to the pass is similar to that of the other valleys. However, beyond the pass the road descends into fantastic elfin and humid temperate forest similar to that described for Canchailloc and San Luis in the Abra Malaga section below, and even farther down into Humid Subtropical Forest which is beyond the scope of this guide.

Key species: Birds mentioned above in the Lakes Piuray and Huaypo and Huacarpay sections are also in these valleys. Additionally: Andean Tinamou, Crested Duck (high lakes) Giant Coot (Lamay valley lakes), Andean and Puna Ibis, Andean Condor, Andean Flicker, Common and Slender-billed Miners, Junín (E), Streak-throated, Scribble-tailed (Yanahuara Valley) and Line-fronted Canasteros, Ground-Tyrants (especially winter), Black-billed Shrike-Tyrant, Rufous-webbed Bush-Tyrant, Crested Becard, Short-tailed Finch, Sierra-Finches, Band-tailed Seedeater, White-winged Diuca-Finch, Bright-rumped Yellow-Finch, Hooded Siskin.

Abra Malaga

This location is still top class birding even though road improvements have cleared much of the roadside vegetation and traffic is faster and more frequent. Two full days, even three, can easily be spent at this locality. The road to Abra Malaga starts in Ollantaytambo at 2800 m (the pass is at 4030 m) and is a modern paved road which continues all the way to Quillabamba and farther into the tropical lowlands.

Ecoan mountain refuge at Abra Malaga

It is a well known spot in birding circles made famous in the late 1970's and early 1980's by pioneer ornithologists John O'Neill and Ted Parker. The true name of this pass is Panti Calla, and this is where, in 1572, mounted armored Spanish Conquistadores marched along its Inca road – still visible on the south side of the valley – to the remote Vilcabamba Mountains to search for and destroy the last rebel Inca stronghold of the Inca Empire. The snow peak that dominates it is known in Quechua as Wakay Wilca ("Sacred Tears"), popularly known today as Veronica, probably named after Saint Veronica. There are three main birding areas detailed below.

ACCESS AND STRATEGY: The starting point for Abra Malaga is the picturesque town of Ollantaytambo 45 km away, which boasts a megalithic Inca fortress and many hotels and restaurants. Two hotels that have nice gardens that attract hummingbirds and other birds are the Pakaritampu Hotel and El Albergue, both close to the railway station and convenient for jumping the train to Machu Picchu. Taxis are available in town and public transport is frequent to Quillabamba. Leave Ollantaytambo on the main road to Quillabamba passing the Inca fortress and initially following the Urubamba River on your left, which then turns north up a spectacular glacial U-shaped hanging valley to the pass. Once heading north with the river at your back you could stop anywhere and be rewarded with good birds. However, there are three main birding areas detailed below – you can bird them in any order and adjust your start time from Ollantaytambo depending on the destination. When it is rainy on the north side of the pass, it is often dry on the south side.

THE PEÑAS AREA (18.8 km and 45 minutes from Ollantaytambo): This semi-humid shrub zone on the south side of the pass around the Collpani Inca ruins [GPS S13 10.614 W072 17.147] 3450 m, and just above at the Peñas ruins at [GPS S13.10.327 W072 17.164] 3537 m to where the shrubs peter out into grassland is best. This habitat can be covered in a few hours. Keep glancing up as Andean Condors are frequent visitors to the valley.

Key species: Taczanowski's Tinamou, Andean Condor, Spot-winged Pigeon, Peruvian Pygmy-Owl, Band-winged Nightjar, Tyrian Metaltail, Shining and White-tufted Sunbeam (E), Giant Hummingbird, White-bellied Hummingbird, Andean Swift, Aplomado Falcon, American Kestrel, Mitred Parakeet (seasonal), Andean Parakeet, Creamy-crested Spinetail (E), Rusty-fronted (E) and Junin (E) Cansteros, Tawny-rumped Tyrannulet, White-crested and Sierran Elaenias, d'Orbigny's, Rufous-breasted and White-

Trail to Polylepis *woodland at Abra Malaga*

browed Chat-Tyrants, Brown-bellied Swallow, Crested Becard, Black-throated Flowerpiercer, Cinerous Conebill, Tit-like Dacnis, Golden-billed Saltator, Peruvian Sierra-Finch, Ash-breasted Sierra-Finch, Greenish Yellow-Finch, Chestnut-breasted Mountain-Finch (E), Black-backed Grosbeak, Hooded Siskin.

THE *POLYLEPIS* AND THE PASS (40 km and one hour +/- from Ollantaytambo): This high elevation site is at 4330 m and prudent acclimatization is recommended before birding here. In the grassland on either side of the pass itself some high Andean puna grassland birds can be seen. A good area is Lagunillas 6.5 km beyond the pass which has a series of small ponds and marshy areas [GPS S13 07.623 W072 19.575] 4030 m. However, the main reason to bird this area is to access the remnant patches of *Polylepis* woodland locally known as Queñua. Fortunately, conservation efforts by the Peruvian NGO Asociación Ecosistemas Andinos (ECOAN) in conjunction with the nearby Quechua community of Tastayoc, affords protection and regeneration of these remnant woodland patches. Do not be surprised if a red ponchoed man in home spun trousers or lady in a multilayered petticoat skirts wielding a small notebook appear from nowhere and asks you for an entrance fee (about 10 Peruvian soles) and for you to sign the book. Be polite and pay up! These humble Andean people are helping protect the *Polylepis* woodland. (Warning – they never have change!) ECOAN, in conjunction with the Tastayoc community has an interpretation centre not far from the *Polylepis* forest. It is located next to the road [GPS S13 08.477 W072 17.942] 4332 m, just 400 m before the pass itself. Basic hut type accommodation is available (10 beds in one room) with all necessary facilities including cooking facilities and toilets. Book through ECOAN: Info@ecoanperu.org & www.ecoanperu.org; phone: 084-227988. A small *Polylepis* grove can be seen and birded 2 km south of the pass on the slope to the east side of the road, but the most extensive area is accessed from near the pass itself, starting at the interpretation centre. From the interpretation centre walk (signposted) toward the ridgetop to the west (you will see a few trees silhouetted against the skyline). It is about a 20 minute walk to the ridge top with lots of stops to catch your breath. Once you get to the ridge you will see a glacial valley with large patches of *Polylepis* trees below you. The patch to your right (north) below steep mossy rock faces is perhaps the best area for Royal Cinclodes. It is steep and you must pick your way through the *Polylepis* groves as best you can as there are no real trails. You have two possible strategies here: bird down the slope and return to your vehicle the way you came (this means a steep climb back up to the ridge) or work the *Polylepis* by making your way down to the valley floor and back to the road at [GPS S13.09.382 W072 16.911] 3805 m – this is 14.1 km by road, but shorter on foot. You can instruct your driver to wait for you there. Obviously, this is not an option if you are driving yourself. If you do not want to do any steep climbs or descents, once on the ridge top, walk left on fairly level trail where you can see some of the *Polylepis* endemics. You need to devote a long morning to this area, especially if you intend to walk down to the valley. The weather can change in a matter of minutes. Be prepared with cold weather and rain gear, sunscreen, snacks and water.

Key species: P = *Polylepis* only. Andean Goose and other waterfowl, Andean Ibis, Variable Hawk, Andean Condor, Mountain Caracara, Gray-breasted Seedsnipe, Blue-mantled Thornbill (P), Andean Hillstar, Stripe-headed Antpitta (P), Puna Tapaculo (P), Line-fronted (P), Streak- throated and Junín Canasteros (E) , Tawny Tit-Spinetail (P), White-browed Tit-Spinetail (P)(E),Plain-breasted Earthcreeper, Royal Cinclodes (P),Cream-winged and White-winged Cinclodes, Ash-breasted Tit-Tyrant (P), d'Orbigny's and Brown-backed Chat-Tyrants, Red-rumped and Rufous-webbed Bush-Tyrants, Correndera, Short-billed and Paramo Pipits, Giant Conebill (P), Tit-like Dacnis, White-winged Diuca-Finch, Bright-rumped Yellow-Finch, Plumbeous Sierra-Finch, Thick-billed Siskin (P).

CANCHAILLOC TO SAN LUIS (68 km and 2 hours from Ollantaytambo): To work this area you need a full day, a packed lunch and a plan to be at Canchailloc [GPS S13 07.218 W072 19.72] ideally no later than 6:30 am. It is a 30 minute drive (13 km) from the pass to Canchailloc. Continue over Abra Malaga pass and stop at Canchailloc at 3708 m and spend the day birding your way downhill to San Luis 27.8 km farther on at 3000 m [GPS S13 04.721 W072 23.426]. There are some small restaurants at Carrizales (5.5 km before San Luis) half way down and some stalls at San Luis where you can get food, but it is best to be self-sufficient and take some drinks and sandwiches. Pay special attention to patches of *Chusquea* bamboo near tree line and do not be in too much of a hurry to get lower. Remnant forest only remains below San Luis and it is not worth going farther. Although not as pristine as it was 20 years ago, it is still excellent for some hard to see elsewhere Peruvian endemics.

Key species: White-rumped and White-throated Hawks, Swallow-tailed Nightjar, Yungas Pygmy-Owl, Andean Snipe, Imperial Snipe, Amethyst-throated Sunangel, Green Violetear, Purple-backed Thornbill, Rufous-capped Thornbill, Tyrian Metlatail, Scaled Metaltail, Sapphire-vented Puffleg, Shining Sunbeam, Violet-throated Starfrontlet, Sword-billed Hummingbird, Great Sapphirewing, Masked Trogon, Gray-breasted Mountain-Toucan, Bar-belied Woodpecker, Crimson-mantled Woodpecker, Golden-plumed Parakeet, Undulated, Rufous (*occobambae* race), Red and White (San Luis), and Rusty-breasted Antpittas, Diademed and Trilling Tapaculos, Streaked Tuftedcheek, Peruvian Treehunter, Pearled Treerunner, Spotted Barbtail, Line-fronted and Junín (E) Canasteros, PunaThistletail, Marcapata Spinetail (E), Sierran and Highland Elaenias, White-throated and White-banded Tyrannulets, Tufted and Unstreaked (E) Tit-Tyrants, Rufous-headed Pygmy-Tyrant, Black-throated Tody-Flycatcher, Cinnamon and Ochraceous-breasted Flycatchers, Smoky and Rufous-bellied Bush-Tyrants, Kalinowski's (Crowned), Golden-browed, Brown-backed and Rufous-breasted Chat-Tyrants, Barred and Band-tailed Fruiteaters, Red-crested Cotinga, Barred Becard, White-collared Jay, Pale-footed and Blue-and-White Swallows, Mountain, Sedge, Fulvous and Inca Wrens, Great Thrush, White-browed (E), Parodi's (E), Supercilliaried, Drab, Three-striped and Black-eared Hemispingus, Rufous-chested, Rust-and-Yellow, Slaty, Blue-capped, Hooded, Grass-green and Golden-collared Tanagers, Lacrimose, Scarlet-bellied, Buff-breasted and Chestnut-bellied Mountain-Tangers, White-browed, Capped and Blue-backed Conebills, Tit-like Dacnis, Mustached, Bluish and Masked Flowerpiercers, Plushcap, Paramo Seedeater, Cuzco (E), and Gray-browed Brushfinches, Yellow-whiskered Bush-Tanager, Spectacled and Slate-throated Whitestarts, Citrine, Russet-crowned and Three-striped Warblers, Southern Mountain and Yellow-billed Caciques, Olivaceous Siskin.

The Inca Trail to Machu Picchu

The Machu Picchu Historical Sanctuary was created to protect the World Heritage Site ruins along the Inca Trail and Machu Picchu itself. It covers elevations from 2650 m at the start to 4200 m at Warmiwanusca (Dead Woman's) Pass and ends at the Machu Picchu ruins at 2425 meters. The Inca Trail is just a tiny part of the massive 30,000 km Incan road network that stretched from Argentina to Colombia. Little did the authorities know that by protecting the trail and archeological sites, a large chunk of intact cloud forest was also being protected.

ACCESS AND STRATEGY: If you are hiking the Inca Trail, the trek begins at Ollantaytambo, kilometer 82 or 88 on the railway line. The trail is only accessible by booking an organized shared or private trek in advance. (Warning: Inca trail Trekking Permits sell out up to 6 months before departure and are strictly limited – book well in advance online via authorized agencies). See www.ManuExpeditons.com. The popular trek itself is around 35 km long (starting at kilometer 82), spectacular, very birdy, a bit overcrowded, tough but worth it. While trail conditions are generally good, and indeed the condition of centuries-old Inca paving stones is astonishing, some steep trail sections require careful footing and good hiking boots with lug soles. Nevertheless, on professionally organized tours, thanks to careful pacing, dedicated guides and professional support staff, the trip is suitable for novices as well as experienced hikers. Porters carry the heavy gear; you walk carrying only your binoculars and a day-pack.

It is a tough hike, most hike it in four days but adding another day is recommended. If birding you could easily spend a week on the trail!

Key species: The trail transverses many habitat types and elevation changes passing through arid montane scrub, semi-humid montane scrub, puna grassland, elfin forest, humid temperate forest and humid pre-montane forest. The list of potential species is too large to list here, just check your elevation and habitat in the annotated checklist and look at the possibilities! One bird you can see between Sayacmaraca and Phuyupatamarca on the trail and not elsewhere is the endemic Vilcabamba Tapaculo. Look for Cuzco Brushfinch (E) between Winay-Wayna and Machu Picchu.

Machu Picchu

The vast majority of foreign visitors who travel to Peru visit Machu Picchu. It is a stunning archaeological site with unbelievable topography and accessible by the Inca Trail (see the above section), by train from Cusco and Ollantaytambo or the backpacker "back door route" alternative via Santa Maria and Santa Theresa.

ACCESS AND STRATEGY: Most folks arrive on the train from Cusco or increasingly more frequently, from Ollantaytambo (Torrent Ducks and White-capped Dippers common along the river and can be seen from the train). There are many trains from Ollantaytambo and a few from Cusco daily – see www.perurail.com & www.incarail.com. There is a more economical route to Machu Picchu via bus from Ollantaytambo to Santa Maria and then shared "colectivos" to Santa Teresa and then local taxis to the hydroelectric plant ("Hidroelectrica"). From here you can hop the twice daily Peru Rail train to Aguas Calientes or walk (3 hours).

The starting point at Machu Picchu is Aguas Calientes at 2070 m. In Aguas Calientes there are many hotels and restaurants (Indio Feliz Bistrotis recommended) to suit all budgets, from cheap and cheerful to elegant luxury. Without doubt the best hotel here for birders is the Machu Picchu Pueblo Hotel owned by Inkaterra (www.inkaterra.com/inkaterra/inkaterra-machu-picchu-pueblo-hotel), it is not the cheapest option but it is the most birder friendly hotel with superb service and onsite naturalist guides who know their birds, plants and wildlife. The grounds of the hotel and nature trails are a birders' paradise. Highly recommended for birders.

THE RUINS: From Aguas Calientes you can take the 20 minute bus journey to the ruins (buy a round trip ticket). Note entrance tickets to Machu Picchu ruins [2425 m] are limited and sometimes sell out in August. You can buy them online, in Cusco or in the town square at Aguas Calientes, but not at the entrance gate. The ruins themselves (a must see!) are not particularly good for birds. But while taking in the site keep your eyes open for birds and for better birding walk along the Inca Trail to Inti Punku (The Sungate) which is better for birds and has stunning vistas of the ruins.

Key species: Black-chested Buzzard-eagle, Black and Chestnut Eagle, Mountain Caracara, White-tipped Swift, Black-tailed Trainbearer, White-winged Black-Tyrant, Tufted Tit-Tyrant, Sierran Elaenia, Blue-and-White Swallow, Spectacled Whitestart, Cinereous Conebill, Slaty and Black-throated Flower-piercers, Blue-capped, Blue-and-yellow and Rust-and-yellow Tanagers, Rufous-collared Sparrow and Tricolored and Cuzco (E) Brushfinches. In the stands of bamboo, the endemic Inca Wren is common and vocal though secretive (a good spot is a bend or two back down the road from the Sanctuary Lodge Hotel near the entrance gate to the ruins).

THE RAILWAY LINE AND THE MANDOR VALLEY: Birding in the forest below the ruins in the Urubamba Valley is far more productive. The area around the Museum, a few bends above the bridge over the Urubamba River (Puente Ruinas), the railway line to the Mandor Valley, the Mandor Valley itself and the garden of the Pueblo Hotel provide exceptionally good birding. On the return from the ruins you might want to ask the driver to let you out a few bends above the river and bird down, taking the side road to the site museum [GPS S13,09 377 W072 32 213]. Do not neglect the museum area – it is good! Check out the Botanical Garden and the Inca terraces on which the museum is built. The walk from Aguas Calientes to the Mandor Valley [1960 m, GPS S13 09 184 W072 32 462] takes about 2 hours birding, from the Puente Ruinas bridge 30 minutes less. To bird the Mandor Valley ("Los Jardines de Mandor") to the waterfall, you will have to pay a small entrance fee of around $4 USD to the landowner. Basic lodging, camping area and meals available – Tel: (+51) 084 634429 / (+51) 940 188155.

Key species: Torrent Duck (river), Fasciated Tiger-Heron (river), Black and Chestnut Eagle, Lyre-tailed Nightjar (near rock faces), Band-bellied Owl, Chestnut-collared and White-tipped Swifts, Green Hermit, Speckled Hummingbird, Long-tailed Sylph, Buff-thighed Puffleg, Collared Inca, Chestnut-breasted Coronet, Booted Racket-tail, White-bellied Woodstar, Green-and-White Hummingbird (E), Masked Trogon, Golden-headed Quetzal, Andean Motmot, Black-streaked Puffbird, Versicolored Barbet (Mandor), Blue-banded Toucanet, Ocellated Piculet, Crimson-bellied (Mandor) and Golden-Olive Woodpeckers, Orange-breasted Falcon, Mitred Parakeet, Scaly-naped Amazon, Variable Antshrike, Scaled Antpitta, Streaked Xenops, Sharp-tailed Streamcreeper, Montane Foliage-gleaner, Spotted Barbtail, Azara's Spinetail, Sclater's, Plumbeous-crowned, Torrent, Ashy-headed, Bolivian, Mottle-cheeked, White-tailed and White-banded Tyrannulets, Streak-necked and Inca (E) Flycatchers, Yellow-olive Tolmomyias, Smoke-colored Pewee, Black Phoebe, Golden-crowned Flycatcher, Dusky-capped Flycatcher, Masked Fruiteater (E) (museum/Mandor), Cock-of-the Rock, Brown-capped Vireo, Green (Inca) Jay, Gray-breasted Wood-Wren, White-capped Dipper (streams), Pale-eyed Thrush, Andean and White-eared Solitaires, Slaty Tanager, Gray headed Bush-Tanager, Oleaginous Hemispingus, Rust-and-Yellow, Silver-beaked, Blue-Gray, Blue-capped, Blue-and-yellow, Yellow-throated, Orange-eared, Fawn-breasted, Saffron-crowned, Beryl–spangled, Yellow-scarfed and Silver-backed Tanagers, Blue Dacnis, Capped Conebill, Masked Flowerpiercer, Dull-colored Grassquit, Yellow-bellied Seedeater, Bananaquit, Chestnut-capped, Gray-browed and Tricolored Brushfinches, Yellow-whiskered and Common Chlorospingus, Tropical Parula, Pale-legged and Russet-crowned Warblers, Dusky-green Oropendola, Orange-bellied Euphonia and Olivaceous Siskin.

AN ANNOTATED CHECKLIST OF THE BIRDS OF MACHU PICCHU AND CUSCO

INCLUDING ABRA MALAGA AND THE SACRED VALLEY OF THE INCAS

E = PERUVIAN ENDEMIC   FP = FORAGING POSITION   S = STATUS

	ENGLISH	SCIENTIFIC	HABITAT	ELEVATION	FP	S
TINAMOUS	**Tinamidae**					
	Hooded Tinamou	*Nothocercus nigrocapillus*	HMF/HPF	2000-3000	T	U
	Brown Tinamou	*Crypturellus obsoletus*	HMF/HPF/SG	< 3000	T	FC
	Taczanowski's Tinamou	*Nothoprocta taczanowskii*	PW/HMS/AA	2800-4000	T	U
	Ornate Tinamou	*Nothoprocta ornata*	PG/AA	3500-4800	T	U
	Andean Tinamou	*Nothoprocta pentlandii*	HMS/AMS/AA	1500-4000	T	FC
DUCKS	**Anatidae**					
	Black-bellied Whistling Duck	*Dendrocygna autumnalis*	FL/FM	3000-3800	W	V
	Andean Goose	*Oressochen melanopterus*	AB/FL/FM	> 3800. Occ 3200	W	FC
	Torrent Duck	*Merganetta armata*	R/S	> 2800	W	C
	Crested Duck	*Lophonetta specularioides*	AB/FL	>3500	W	FC
	Yellow-billed Teal	*Anas flavirostris*	R/AB/FL/FM	2500-4500	W	C
	Yellow-billed Pintail	*Anas georgica*	FL/FM	> 3000	W	C
	White-cheeked Pintail	*Anas bahamensis*	FL/FM	3000	W	U
	Puna Teal	*Anas puna*	FL/FM	> 3000	W	C
	Blue-winged Teal	*Anas discors*	FL/FM	3000	W	bm U
	Cinnamon Teal	*Anas cyanoptera*	FL/FM	> 3000	W	FC
	Red Shoveler	*Anas platalea*	FL/FM	3000	W	am V
	Andean (Ruddy) Duck	*Oxyura (jamaicensis) ferruginea*	FL/FM	> 3000	W	FC
GUANS	**Cracidae**					
	Sickle-winged Guan	*Chamaepetes goudotii*	HPF/HMF	1800-2500	C	R
	Andean Guan	*Penelope montagnii*	HPF/HMF	1800-3500	M/C	FC
NEW WORLD QUAIL	**Odontophoridae**					
	Rufous-breasted Wood-Quail	*Odontophorus speciosus*	HPF	1800-2600	T	FC
	Stripe-faced Wood-Quail	*Odontophorus balliviani*	HMF/HPF	1800-3300	T	U
GREBES	**Podicipedidae**					
	White-tufted Grebe	*Rollandia rolland*	FL/FM	3400-4500	W	FC
	Silvery Grebe	*Podiceps occipitalis juninensis*	FL	3000-5000	W	U

FLAMINGOS Phoenicopteridae					
Chilean Flamingo	*Phoenicopterus chilensis*	FL/FM	>3000	W	R
STORKS Ciconidae					
Jabiru	*Jabiru mycteria*	R/FL/FM	< 3300	T/W	V
Wood Stork	*Mycteria americana*	R/FL/FM	< 3300	T/W	V
CORMORANTS Phalacrocoracidae					
Neotropic Cormorant	*Phalacrocorax brasilianus*	R/FL	2000- 4200	W	FC
HERONS Ardeidae					
Fasciated Tiger-Heron	*Tigrisoma fasciatum*	R/S	> 2200 Vag. to 3300	T/W	U
Black-crowned Night-heron	*Nycticorax nycticorax*	AB/FL/FM	3000-4700	T/W	FC
Yellow-crowned Night-heron	*Nyctanassa violacea*	AB/FL/FM	Vag. 3200	T/W	V
Striated Heron	*Butorides striatus*	FL/FM	Vag. <4000	T/W	U
Cattle Egret	*Bubulcus ibis*	FL/FM	< 3200 Vag. to 4300	T/W	FC
Cocoi Heron	*Ardea cocoi*	R/FL	< 3500	T/W	R
Great Egret	*Ardea alba*	R/AB/FL/FM	< 4000	T/W	FC
Snowy Egret	*Egretta thula*	R/AB/FL/FM	<2 600 Vag. to 4000	T/W	FC
Little Blue Heron	*Egretta caerulea*	FL/FM	3000	T/W	FC
IBIS Threskiornithidae					
Puna Ibis	*Plegadis ridgwayi*	PG/AB/FL/FM	3000-4800	T/W	C
Andean Ibis	*Theristicus branickii*	PG/AB	3500-4700	T	FC
AMERICAN VULTURES Cathartidae					
Turkey Vulture	*Cathartes aura*	R/SG/AA/RT	< 2500 Vag to 4000	T/A	R
Andean Condor	*Vultur gryphus*	PG/AMS	> 2500	T/A	U
OSPREYS Pandionidae					
Osprey	*Pandion haliaetus*	FL	3000	A/W	bm R
HAWKS Accipitridae					
Swallow-tailed Kite	*Elanoides forficatus*	HPF	< 2400 Vag.to 4000	C/A	U
Black-and-chestnut Eagle	*Spizaetus isidori*	HMF/HPF	1800-3500	C/A	U
Cinereous Harrier	*Circus cinereus*	PG/AA/FM	2500-4500	T/A	FC
Semicollared Hawk	*Accipiter collaris*	HPF	< 2500	C	U
Plain-breasted Hawk	*Accipiter ventralis*	SG/HMF/HPF	1800-3500	C	U
Montane Solitary Eagle	*Buteogallus solitarius*	HPF	< 2200	C	R
Roadside Hawk	*Rupornis magnirostris*	R/SG/RT	< 2600	C	U
White-rumped Hawk	*Parabuteo leucorrhous*	HPF	1800-3700	C	U

Variable Hawk	*Geranoaetus polyosoma*	>	1800-4600	T/C	FC
Black-chested Buzzard-Eagle	*Geranoaetus melanoleucus*	>	1800-3800	T/A	C
Broad-winged Hawk	*Buteo platypterus*	>	< 3000	C	bm U
White-throated Hawk	*Buteo albigula*	HMF/HPF	1800-3300	C	FC

RAILS Rallidae

Ocellated Crake	*Micropygia schomburgkii*	?	Vag.	T	V
Paint-billed Crake	*Mustelirallus erythrops*	FL/FM	3000-3400	T	V/R
Spotted Rail	*Pardirallus maculatus*	FL/FM	3000	T	V
Plumbeous Rail	*Pardirallus sanguinolentus*	FL/FM	< 4000	T	FC
Common Gallinule	*Gallinula galeata*	FL/FM	2200-4400	T/W	C
Purple Gallinule	*Porphyrio martinica*	FL/FM	< 3200 Vag	T	R
Giant Coot	*Fulica gigantea*	FL	3900 - 4600	T/W	U
Slate-colored Coot	*Fulica ardesiaca*	FL	3000-4700	T/W	C

SUNBITTERNS Eurypygidae

Sunbittern	*Eurypyga helias*	R/FL	2000	T	V

PLOVERS Charadriidae

American Golden Plover	*Pluvialis dominica*	FL/FM	3000-4500	T	U
Black-bellied (Grey) Plover	*Pluvialis squatarola*	FL/FM	3000	T	R
Andean Lapwing	*Vanellus resplendens*	PG/AB/FL/FM	3000-4600	T	C
Semi-palmated Plover	*Charadrius semipalmatus*	FL/FM	3000	T	bm U
Collared Plover	*Charadrius collaris*	FL/FM	3000-4500	T	U
Puna Plover	*Charadrius alticola*	PG/AB/FL	3000-4500	T	U
Diademed Sandpiper-Plover	*Phegornis mitchellii*	PB/AB/S	4100-5000	T	R

STILTS Recurvirostridae

White-backed (Black-necked) Stilt	*Himantopus (mexicanus) melanurus*	R/FL/FM	> 3000	T	FC
Andean Avocet	*Recurvirostra andina*	FL/FM	> 3000	T	R

SANDPIPERS AND SNIPES Scolopacidae

Upland Sandpiper	*Bartramia longicauda*	FL/FM	3000-3500	T	bm U
Whimbrel	*Numenius phaeopus*	FL/FM	3000	T	bm V
Hudsonian Godwit	*Limosa haemastica*	FL/FM	3000-3500	T	bm R
Sanderling	*Calidris alba*	FL/FM	3000	T	bm R
Semi-palmated Sandpiper	*Calidris pusilla*	FL/FM	3000	T	bm R
Least Sandpiper	*Calidris minutilla*	FL/FM	3000	T	bm U
White-rumped Sandpiper	*Calidris fuscicollis*	FL/FM	3000	T	bm V

Baird's Sandpiper	*Calidris bairdii*	FL/FM	3000-4500	T	bm FC
Pectoral Sandpiper	*Calidris melanotos*	FL/FM	3000-4500	T	bm U
Stilt Sandpiper	*Calidris himantopus*	FL/FM	3000-3500	T	bm U
Buff-breasted Sandpiper	*Tryngites subruficollis*	FL/FM	3000-3500	T	bm R
Imperial Snipe	*Gallinago imperialis*	EF/PG	2745-3500	T	R
Andean Snipe	*Gallinago jamesoni*	EF/PG	3000-3800	T	U
Wilson's Phalarope	*Phalaropus tricolor*	FL/FM	3000-4500	T/W	bm FC
Spotted Sandpiper	*Actitis macularia*	R/S	> 1800	T	bm C
Solitary Sandpiper	*Tringa solitaria*	R/FL/FM	< 4000	T	bm U
Greater Yellowlegs	*Tringa melanoleuca*	R/FL/FM	< 4500	T	bm FC
Lesser Yellowlegs	*Tringa flavipes*	R/FL/FM	< 4500	T	bm FC
SEEDSNIPES Thinocoridae					
Rufous-bellied Seedsnipe	*Attagis gayi*	PG/AB	4000-5000	T	U
Gray-breasted Seedsnipe	*Thinocorus orbignyianus*	PG	3500-5000	T	FC
GULLS AND TERNS Laridae					
Andean Gull	*Chroicocephalus serranus*	R/AB/FL/FM	2000-4500	T/W	C
Laughing Gull	*Leucophaeus atricilla*	R/AB/FL/FM	< 3800	T/W	bm R
Franklins Gull	*Leucophaeus pipixcan*	R/AB/FL/FM	< 3000	T/W	bm V
Large-billed Tern	*Phaetusa simplex*	R/AB/FL/FM	< 3000	T/W	V
Black Tern	*Childonias niger*	R/AB/FL/FM	< 3000	T/W	bm V
Common Tern	*Sterna hirundo*	R/AB/FL/FM	< 3000	T/W	bm V
SKIMMERS Rhyncopidae					
Black Skimmer	*Rynchops niger*	FL	3300-3700	T/W	R
PIGEONS AND DOVES Columbidae					
● Rock Pigeon	*Columba livia*	>	< 2000	T/C	C
Spot-winged Pigeon	*Patagioenas maculosa*	HMS/AMS/AA	2000-4000	C	C
Band-tailed Pigeon	*Patagioenas fasciata*	HMS/HMF/HPF	2000-3000	C	FC
Plumbeous Pigeon	*Patagioenas plumbea*	HPF	< 2000	C	R
Ruddy Quail-Dove	*Geotrygon montana*	HPF	< 2000	C	R
White-tipped Dove	*Leptotila verreauxi*	SG/HPF	< 2500	T/U	FC
White-throated Quail-Dove	*Zentrygon frenata*	HMF/HPF/SG	1800-3000	T	U
Eared Dove	*Zenaida auriculata*	AMS/AA/RT	2500-4000	T/M	C
Maroon-chested Ground-Dove	*Claravis mondetoura*	HMF/HPF/B	1800-3000	T/M	N/U

Bare-faced Ground-Dove	*Metriopelia ceciliae*	PG/AMS	2500-4500	T	C
Black-winged Ground-Dove	*Metriopelia melanoptera*	PW/PG/AMS	2000-4300	T	FC

CUCKOOS AND ANIS   Cuculidae

Squirrel Cuckoo	*Piaya cayana*	SG/HPF	< 2500	M/C	R
Dark-billed Cuckoo	*Coccyzus melacoryphus*	>	< 2400.Vag. to 3400	M/C	U
Yellow-billed Cuckoo	*Coccyzus americanus*	>	Vag 2000	M/U	bm V
Greater Ani	*Crotophaga major*	RT/FM/SG	Vag 2000	M/C	V
Smooth-billed Ani	*Crotophaga ani*	SG/AA/RT	< 2800	T/C	R

BARN OWLS   Tytonidae

Barn Owl	*Tyto alba*	>	1800-4000	C	U

TYPICAL OWLS   Strigidae

White-throated Screech-Owl	*Megascops albogularis*	HMF/HPF	2000-3000	C	U
Band-bellied Owl	*Pulsatrix melanota*	HPF	2000	C	U
Great (Lesser) Horned Owl	*Bubo (magellanicus) virginianus*	PG/AMS	> 3000	C	U
Rufous-banded Owl	*Ciccaba albitarsus*	HMF/HPF	1900-3500	C	FC
Yungas Pygmy-Owl	*Glaucidium bolivianum*	HMF/HPF	1800-3000	M/C	FC
Peruvian Pygmy-Owl	*Glaucidium peruanum*	AMS/AA	< 3400	M/C	U
Burrowing Owl	*Athene cunicularia*	PG/AMS	3000-4000	T	FC

OILBIRD   Steatornithidae

Oilbird	*Steatornis caripensis*	HPF	< 3000	C/A	R

NIGHTHAWKS AND NIGHTJARS   Caprimulgidae

Nacunda Nighthawk	*Chordeilis nacunda*	FL/FM/AA	3000	A	am V
Rufous-bellied Nighthawk	*Lurocalis rufiventris*	HPF	1800-3300	A	U
Band-winged Nightjar	*Systellura longirostris*	PW/PG/HMS/AMS	2400-4200	A/T	FC
Swallow-tailed Nightjar	*Uropsalis segmentata*	EF/HMF	2500-3600	A/T	U
Lyre-tailed Nightjar	*Uropsalis lyra*	HMF/HPF	1800-2100	A/T	U

SWIFTS   Apodidae

Chestnut-collared Swift	*Streptoprocne rutilus*	HPF/HMF	< 3000	A	U
White-collared Swift	*Streptoprocne zonaris*	<	1800- 4000	A	FC
Chimney Swift	*Chaetura pelagica*	AMS/AA	< 3300	A	bm U
White-tipped Swift	*Aeronautes montivagus*	HMF/HPF/HMS	1800-2400	A	FC
Andean Swift	*Aeronautes andecolus*	AMS/AA	2500-4500	A	C

HUMMINGBIRDS Trochilidae

Buff-tailed Sicklebill	*Eutoxeres condamini*	HMF/HPF	< 2500 Vag.to 3300	U	R
Green Hermit	*Phaethornis guy*	HPF	< 2300	U	U
Green-fronted Lancebill	*Doryfera ludovicae*	HPF	1800-2500	U/M	U
Green Violet-ear	*Colibri thalassinus*	SG/HMS	1800-2900	M/C	FC
Sparkling Violet-ear	*Colibri coruscans*	<	1800-4000	U/C	C
Amethyst-throated Sunangel	*Heliangelus amethysticollis*	HMF/HPF	2000-3700	U/M	C
Speckled Hummingbird	*Adelomyia melanogenys*	HPF	1800-2800	U	C
Long-tailed Sylph	*Aglaiocercus kingi*	HMF/HPF	1800-2700	M/C	C
Andean Hillstar	*Oreotrochilus estella*	PW/PG/AMS	3500-4500	T/U	FC
Black-tailed Trainbearer	*Lesbia victoriae*	PW/AMS/AA	2700-4000	U/C	FC
Green-tailed Trainbearer	*Lesbia nuna*	SG/HMS/AMS/AA	2000-3800	U/C	FC
Purple-backed Thornbill	*Ramphomicron microrhynchum*	EF/HMS	2500-3650	M/C	U
Rufous-capped Thornbill	*Chalcostigma ruficeps*	HMF/HPF	1800-3600	U/M	U
Olivaceous Thornbill	*Chalcostigma olivaceum*	PG/PW	4000-4600	T/U	U
Blue-mantled Thornbill	*Chalcostigma stanleyi*	PW/PG	3600-4400	U/C	FC
E Bearded Mountaineer	*Oreonympha nobilis*	HMS/AMS/AA	2700-3900	U/M	FC
Tyrian Metaltail	*Metallura tyrianthina*	SG/EF/HMF/HMS	2500-3600	U/M	C
Scaled Metaltail	*Metallura aeneocauda*	EF/HMS	2800-3600	U/C	U
Buff-thighed Puffleg	*Haplophaedia assimilis*	HPF	1800-2500	U	U
Sapphire-vented Puffleg	*Eriocnemis luciani*	HMF/HPF	2400-3500	U/M	FC
Shining Sunbeam	*Aglaeactis cupripennis*	PW/HMS	2500-4600	U/C	FC
E White-tufted Sunbeam	*Aglaeactis castelnaudii*	PW/HMS	2600-4100	U/C	FC
Bronzy Inca	*Coeligena coeligena*	HPF	1800-2200	U/M/C	U
❦ Collared Inca	*Coeligena torquata*	HMF/HPF	1800-3000	U/M	C
Violet-throated Starfrontlet	*Coeligena violifer*	EF/HMF	2500-3900	U/M	C
Mountain Velvetbreast	*Lafresnaya lafresnayi*	HMF/HPF	1800-3200	U	U
Sword-billed Hummingbird	*Ensifera ensifera*	EF/HMF/HMS/VT	2400-3600	U/C	FC

Great Sapphirewing	*Pterophanes cyanopterus*	EF/HMF/PW	2600-3700	U/M	FC
Chestnut-breasted Coronet	*Boissonneaua matthewsii*	HPF	1800-3300	M/C	FC
Booted Racket-tail	*Ocreatus underwoodii*	HMF/HPF	1800-2400	U	FC
Fawn-breasted Brilliant	*Heliodoxa rubinoides*	HPF	2000	M/C	U
Giant Hummingbird	*Patagona gigas*	HMS/AMS/AA	2500-4000	U/C	C
White-bellied Woodstar	*Chaetocercus mulsant*	SG/HMF/HPF	1800-3000	U/C	FC
Swallow-tailed Hummingbird	*Eupetonema macroura*	SG/AA/AMS	1800	U/M	R
White-bellied Hummingbird	*Amazilia chionogaster*	HMS/AMS/AA	1800-3400	U/C	FC
Green-and-white Hummingbird	*Amazilia viridicauda*	SG/HPF	1800-2800	U/C	C
TROGONS AND QUETZALS Trogonidae					
Golden-headed Quetzal	*Pharomachrus auriceps*	HPF	1800-2300	M/C	U
Crested Quetzal	*Pharomachrus antisianus*	HPF	1800-2300	M/C	U
Masked Trogon	*Trogon personatus*	HMF/HPF	1800-3300	M	FC
MOTMOTS Momotidae					
Rufous Motmot	*Baryphthengus martii*	HPF	2000	U/M	V?
Andean Motmot	*Momotus aequatorialis*	SG/HPF	1500-2400	U/M	C
PUFFBIRDS Bucconidae					
Black-streaked Puffbird	*Malacoptila fulvogularis*	HPF	1800-2000	U	U
NEW WORLD BARBETS Capitonidae					
Versicolored Barbet	*Eubucco versicolor*	HPF	<2100	C	U
TOUCANS					
Black-throated (Emerald) Toucanet	*Aulacorhynchus (prasinus) atrogularis*	HPF	<2000	C	R
Blue-banded Toucanet	*Aulacorhynchus coeruleicinctis*	HPF	1800-2500	C	U
Gray-breasted Mountain-Toucan	*Andigena hypoglauca*	EF/HMF/HPF	2000-3500	C	FC
WOODPECKERS Picidae					
Ocellated Piculet	*Picumnus dorbygnianus*	HMS/HPF	<2500	M/C	FC
Bar-bellied Woodpecker	*Veniliornis nigriceps*	EF/HMF/B	2400-3500	M/C	FC
Golden-olive Woodpecker	*Piculus rubiginosus*	SG/HMF/MPF	<2400	C	FC

Crimson-mantled Woodpecker	*Piculus rivolii*	EF/HMF/HPF	1000-3400	M/C	FC
Andean Flicker	*Colaptes rupicola*	PG/AMS	2500-4500	T	C
Crimson-bellied Woodpecker	*Campephilus haematogaster*	HPF	< 2300	U/M	U

FALCONS Falconidae

Mountain Caracara	*Phalcoboenus megalopterus*	PG/AMS	3500-5000	T	C
American Kestrel	*Falco sparverius*	>	2500-4500	T/C	C
Orange-breasted Falcon	*Falco deiroleucus*	R/HPF	< 2100	T/A	U
Aplomado Falcon	*Falco femoralis*	PG/AMS/AA	2000-4600	T/C	FC
Peregrine Falcon	*Falco peregrinus*	HMS/AMS/PG	> 2000	T/A	U

PARROTS Psittacidae

Barred Parakeet	*Bolborhynchus lineola*	HMF/HPF/B	< 3300	T/C	N/U
Andean Parakeet	*Bolborhynchus orbygnesius*	HMS/AA/RT	2400-3900	C	FC
Speckle-faced Parrot	*Pionus tumultuosus*	HMF/HPF	1800-3300	C	FC
Scaly-naped Amazon	*Amazona mercenaria*	HMF/HPF	1800-3500	C	U
Golden-plumed Parakeet	*Leptosittaca branickii*	HMF/HPF	2700-3500	C	U
Mitred Parakeet	*Psittacara mitrata*	HMS/HPF	1800-3300	C	C

ANTBIRDS Thamnophilidae

Variable Antshrike	*Thamnophilus caerulescens*	SG/HPF	< 2400	U/M	FC
Streak-headed Antbird	*Drymophila striaticeps*	HMF/HPF/B	< 2500	U/M	U

ANTPITTAS Grallaridae

Undulated Antpitta	*Grallaria squamigera*	EF/HMF	2000-3500	T	FC
Scaled Antpitta	*Grallaria guatimalensis*	HPF	< 2000	T	R
Stripe-headed Antpitta	*Grallaria andicolus punensis*	PW/EF	> 3000	T	FC
E Red-and-white Antpitta	*Grallaria erythroleuca*	HMF/HPF/B	2000-3000	T	U
Rufous Antpitta	*Grallaria rufula occabambae*	EF/HMF/B	2600-3600	T	FC
Rusty-breasted Antpitta	*Grallaricula ferrugineipectus leymebambae*	HMF/HPF	2000-3300	U	R

TAPACULOS Rhinocryptidae

Trilling Tapaculo	*Scytalopus parvirostris*	HMF/HPF	2100-3600	U	U
E Vilcabamba Tapaculo	*Scytalopus urubambae*	PG/EF/HMF	3500-4100	U/T	FC

Puna Tapaculo	*Scytalopus simonsi*	PG/EF/PW/HMF	2900-4300	U/T	FC
Diademed Tapaculo	*Scytalopus schulenbergi*	EF/HMF/B	2750-3400	U	FC

ANTHRUSHES   Formicariidae

Rufous-breasted Antthrush	*Formicarius rufipectus*	HPF	< 2400	T	U
Barred Antthrush	*Chamaeza mollissima*	HMF/HPF	2000-3000	T	U

OVENBIRDS   Furnariidae

Slender-billed Miner	*Geositta tenuirostris*	PG/AA	>3000	T	FC
Common Miner	*Geositta cunicularia*	PG	> 3500	T	FC
Strong-billed Woodcreeper	*Xiphocolaptes promeropirhynchus*	HPF	< 3100	M/C	R
Greater Scythebill	*Drymotoxeres pucherani*	HMF	3650	M/C	R
Montane Woodcreeper	*Lepidocolaptes lacrymiger*	HPF	< 3200	M	FC
Streaked Xenops	*Xenops rutilans*	SG/HPF	< 2500	U/M	FC
Streaked Tuftedcheek	*Pseudocolaptes boissonneautii*	EF/HMF	2300-3400	M/C	FC
Rusty-winged Barbtail	*Premnornis guttuliger*	HPF	< 2500	U	U
Sharp-tailed Streamcreeper	*Lochmias nematura*	R/S	1800-2300	T	R
Wren-like Rushbird	*Phleocryptes melanops*	FL/FM	3200-4200	U	FC
Buff-breasted Earthcreeper	*Upucerthia validirostris*	PG	> 3400	T	FC
Cream-winged Cinclodes	*Cinclodes albiventris*	PG/AB/HMS/S	> 2750	T	C
Royal Cinclodes	*Cinclodes aricomae*	PW	3700-4600	T	R
White-winged Cinclodes	*Cinclodes atacamensis*	PG/AB/HMS/S	2800-4600	T	FC
Montane Foliage-gleaner	*Anabacerthia striaticollis*	HPF	< 2100	M	U
Rufous-backed (Peruvian) Treehunter	*Thripadectes scrutator*	EF/HMF	2100-3500	U	U
Striped Treehunter	*Thripadectes holostictus*	HPF	< 2400	U	U
Spotted Barbtail	*Premnoplex brunnescens*	HPF	< 2800	U	U
Pearled Treerunner	*Margarornis squamiger*	EF/HMF	2100-3700	M	C
Tawny Tit-Spinetail	*Leptasthenura yanacensis*	PW	3900-4500	U/C	U
E White-browed Tit-Spinetail	*Leptasthenura xenothorax*	PW	3800-4500	U/C	R

	Andean Tit-Spinetail	*Leptasthenura andicola*	PW/HMS	3500-4200	U/C	R
	Streak-fronted Thornbird	*Phacellodomus striaticeps*	AMS/HMS	2900-3800	U/C	U
	Line-fronted Canastero	*Asthenes urubambensis*	EF/PW	3100-3800	U	FC
E	Junin Canastero	*Asthenes virgata*	PG/EF/HMS	3350-4500	T/U	U
	Scribble-tailed Canastero	*Asthenes maculicauda*	PG/EF/HMS	3000-3900	T/U	R
	Streak-backed Canastero	*Asthenes wyatti*	PG	3500-4600	T	R
	Streak-throated Canastero	*Asthenes humilis*	PG	3700-4800	T	C
	Cordilleran Canastero	*Asthenes modesta*	PG	3600-4600	T	U
	Puna Thistletail	*Asthenes helleri*	EF/HMF	2700-3700	U	FC
E	Rusty-fronted Canastero	*Asthenes ottonis*	HMS/AMS/AA	2900-4000	U	C
E	Marcapata Spinetail	*Cranioleuca marcapatae*	EF/HMF/B	2500-3350	U	FC
E	Creamy-crested Spinetail	*Cranioleuca albicapilla*	HMS/AA	2500-3600	U/M	FC
	Azara's Spinetail	*Synallaxis azarae*	HMF/HPF	< 3100	U	C

TYRANT FLYCATCHERS Tyrannidae

	Sclater's Tyrannulet	*Phyllomyias sclateri*	SG/HPF	< 2200	C	FC
	Black-capped Tyrannulet	*Phyllomyias nigrocapillus*	HMF/HPF	1800-2100	C	U
	Ashy-headed Tyrannulet	*Phyllomyias cinereiceps*	HMF/HPF	< 2400	C	U
	Tawny-rumped Tyrannulet	*Phyllomyias uropygialis*	SG/HMF/HPF/HMS	2400-3300	C	U
	Plumbeous-crowned Tyrannulet	*Phyllomyias plumbeiceps*	HPF	< 2200	C	FC
	Yellow-bellied Elaenia	*Elaenia flavogaster*	SG/AA/RT	< 2000	C	R
	White-crested Elaenia	*Elaenia albiceps*	SG/HMS/AMS/RT	1800-3300	C	C
	Small-billed Elaenia	*Elaenia parvirostris*	SG/HMS	< 3000	C	bm R
	Highland Elaenia	*Elaenia obscura*	SG/HPF	1800-2700	M/C	U
	Sierran Elaenia	*Elaenia pallatangae*	SG/HMF/HPF	1800-3500	C	C
	White-tailed Tyrannulet	*Mecocerculus poecilocercus*	HPF	1800-2500	C	U
	White-banded Tyrannulet	*Mecocerculus stictopterus*	HMF/HPF	2400-3600	C	FC
	White-throated Tyrannulet	*Mecocerculus leucophrys*	EF/PW/HMF	2500-4600	C	C
	Ash-breasted Tit-Tyrant	*Anairetes alpinus*	PW	3700-4600	M/C	U

	Common Name	Scientific Name	Habitat	Elevation	Abundance	Status
	Yellow-billed Tit-Tyrant	*Anairetes flavirostris*	HMS/AMS/RT	1800-3600	U/C	FC
	Tufted Tit-Tyrant	*Anairetes parulus*	HMF/HMS	1800-3700	U/C	FC
E	Unstreaked Tit-Tyrant	*Uromyias agraphia*	EF/HMF/B	2700-3400	U	U
	Torrent Tyrannulet	*Serpophaga cinerea*	R/S	1800-3000	T/U	C
	Subtropical Doradito	*Pseudocolopteryx acutipennis*	FM/AA	3000-3200	U	U
	Rufous-headed Pygmy-Tyrant	*Pseudotriccus ruficeps*	HMF/HPF	2000-3350	U	FC
	Bolivian Tyrannulet	*Zimmerius bolivianus*	HPF	1800-2600	C	FC
	Marble-faced Bristle-Tyrant	*Phylloscartes opthalmicus*	HPF	< 2000	M/C	R
	Mottle-cheeked Tyrannulet	*Phylloscartes ventralis*	HPF	< 2400	C	FC
	Streak-necked Flycatcher	*Mionectes striaticollis*	HMF/HPF	1800- 3000	U/M	C
	Slaty-capped Flycatcher	*Leptopogon superciliaris*	HPF	< 2000	U/M	U
E	Inca Flycatcher	*Leptopogon taczanowskii*	HMF/HPF	1800-2800	U/M	FC
	Many-colored Rush-tyrant	*Tachuris rubigastra*	FM/FL	< 4200	U	FC
	Scale-crested Pygmy-Tyrant	*Lophotriccus pileatus*	HPF	< 2200	U	R
	Black-throated Tody-Tyrant	*Hemitriccus granadensis*	HMF/HPF	2000-3100	U	U
	Common Tody-Flycatcher	*Todirostrum cinereum*	SG/RT/T	<2000	U/C	FC
	Yellow-Olive Tolmomyias	*Tolmomyias sulphurescens*	SG/HPF	<2000	M/C	U
	Bran-colored Flycatcher	*Myiophobus fasciatus*	SG/RT	<2000	U	am R
	Cinnamon Flycatcher	*Pyrrhomyias cinnamomeus*	SG/HMF/HPF	1800-3400	C	C
	Cliff Flycatcher	*Hirundinea ferruginea*	SG/HPF	< 2200	C	U
	Ochraceous-breasted Flycatcher	*Nephelomyias ochraceiventris*	EF/HMF	2200-3700	M/C	FC
	Alder Flycatcher	*Empidonax alnorum*	SG/HMF/HPF	< 3400	C	bm R
	Olive-sided Flycatcher	*Contopus cooperi*	SG/HMF/HPF	< 2100	C	bm U
	Smoke-colored Pewee	*Contopus fumigatus*	SG/HMF/HPF	< 2800	C	FC
	Western Wood-Pewee	*Contopus soridulus*	SG/HMF/HPF	< 2 000	C	bm U
	Olive Flycatcher	*Mitrephanes olivaceus*	HPF	< 2000	U/M	R
	Black Phoebe	*Sayornis nigricans*	R/S	< 2800	T/C	C
	Vermillion Flycatcher	*Pyrocephalus rubinus*	SG/AA	< 2000	T/C	am R
	Andean Negrito	*Lessonia oreas*	FL/AB	3100-4600	T	U

Rufous-tailed Tyrant	*Knipolegus poecilurus*	SG/HPF/HMS	1800-2300	U/M	R	
Plumbeous Black-Tyrant	*Kniploegus cabanisi*	HMF/HPF	1800-2700	U	R	
White-winged Black-Tyrant	*Knipolegus aterrimus*	SG/HMF/HPF/RT	< 3000	C	C	
Spectacled Tyrant	*Hymenops perspicillaus*	FM/FL	3400	T	am V	
Little Ground-Tyrant	*Muscisaxicola fluviatilis*	AA	2000	T/U	R	
Spot-billed Ground-Tyrant	*Muscisaxicola maculirostris*	AMS/AA	2000-4000	T	FC	
Taczanowski's Ground-Tyrant	*Muscisaxicola griseus*	PG/AMS	3300-4800	T	FC	
Puna Ground-Tyrant	*Muscisaxicola juninensis*	PG/AB	3800-4800	T	FC	
Cinereous Ground-Tyrant	*Muscisaxicola cinereus*	PG	4000-4500	T	am U	
White-fronted Ground-Tyrant	*Muscisaxicola albifrons*	PG/AB	3800-4800	T	U	
Ochre-naped Ground-Tyrant	*Muscisaxicola flavinucha*	PG	3800-4700	T	am U	
Rufous-naped Ground-Tyrant	*Muscisaxicola rufivertex*	PG/AMS	3000-4500	T	FC	
White-browed Ground-Tyrant	*Muscisaxicola albilora*	PG	3000-4200	T	am U	
Black-billed Shrike-Tyrant	*Agriornis montanus*	PG/AMS/AA	3000-4500	T/U	FC	
White-tailed Shrike-Tyrant	*Agriornis albicauda*	PG/AMS/AA	3000-4300	T/U	R	
Gray-bellied Shrike-Tyrant	*Agriornis micropterus*	PG/AMS/AA	3000-4100	T/U	am R	
Streak-throated Bush-Tyrant	*Myiotheretes striaticollis*	PW/PG/AMS/HMS	2000-3700	C	FC	
Smoky Bush-Tyrant	*Myiotheretes fumigatus*	HMF/HPF	2300-3400	C	U	
Rufous-bellied Bush-Tyrant	*Myiotheretes fuscorufus*	HMF/HPF	2000-3400	C	U	
Red-rumped Bush-Tyrant	*Cnemarchus erythropygius*	EF/PW/PG	3000-4300	T/C	U	
Rufous-webbed Bush-Tyrant	*Polioxolmis rufipennis*	PW/AMS/AA	3000-4600	T/U	U	
Crowned (Kalinowski's) Chat-Tyrant	*Ochthoeca (spodionota) frontalis*	EF/HMF/HPF	2800-3700	U	FC	
Golden-browed Chat-Tyrant	*Ochthoeca pulchella*	HMF/HPF	1800-3100	U	U	
Slaty-backed (Maroon-belted) Chat-Tyrant	*Ochthoeca (thoracica) cinnamomeiventris*	HMF/HPF/S	1800-3300	U/M	FC	

Common Name	Scientific Name	Habitat	Elevation		
Rufous-breasted Chat-Tyrant	*Ochthoeca rufipectoralis*	EF/HMF/HMS	2300-4000	U/C	C
Brown-backed Chat-Tyrant	*Ochthoeca fumicolor*	PW/EF/HMF	2500-4200	T/C	FC
D'Orbigny's Chat-Tyrant	*Ochthoeca oenanthoides*	EF/HMS/AMS	3200-4400	T/C	U
White-browed Chat-Tyrant	*Ochthoeca leucophrys*	HMS/AMS	2400-4200	T/C	C
Piratic Flycatcher	*Legatus leucophaius*	SG/HPF	< 2000	M/C	R
Social Flycatcher	*Myiozetetes similis*	SG/HPF	< 2000	M/C	FC
Great Kiskadee	*Pitangus sulphuratus*	SG/HMF/HPF	< 2000	M/C	V
Lemon-browed Flycatcher	*Conopias cinchoneti*	HPF	< 2000	C	U
Golden-crowned Flycatcher	*Myiodynastes chrysocephalus*	SG/HPF	< 2700	M/C	C
Streaked Flycatcher	*Myiodynastes maculatus*	RT/SG/HPF	< 2000	M/C	R
Crowned Slaty-Flycatcher	*Empidonomus aurantioatrocristatus*	HMS/HMF/HPF	3200	C	am V
Tropical Kingbird	*Tyrannus melancholicus*	SG/R/RT	<2500	C	FC
Fork-tailed Flycatcher	*Tyranus savana*	AA/SG	3000	M/C	am R
Eastern Kingbird	*Tyrannus tyrannus*	>	1800 -3000	C	bm U
Rufous Casiornis	*Casiornis rufa*	SG/HPF	< 2000	M/C	am R
Dusky-capped Flycatcher	*Myiarchus tuberculifer*	SG/HMF/HPF	< 3250	M/C	U
Pale-edged Flycatcher	*Myiarchus cephalotes*	SG/HMF/HPF	< 2600	M/C	U

COTINGAS Cotingidae

Band-tailed Fruiteater	*Pipreola intermedia*	HMF/HPF	2000-3000	U/M	U
Barred Fruiteater	*Pipreola arcuata*	HMF/HPF	2100-3500	M/C	FC
E Masked Fruiteater	*Pipreola pulchra*	HPF	1800 - 2400	C	FC
Red-crested Cotinga	*Ampelion rubrocristata*	EF/HMF	2400-3700	C	C
Chestnut-crested Cotinga	*Ampelion rufaxilla*	HMF/HPF	1800-2700	C	R
Andean Cock-of-the-rock	*Rupicola peruviana*	HPF	< 2400	C	FC

BECARDS Tityridae

Barred Becard	*Pachyramphus versicolor*	HMF/HPF	1800-3000	C	FC
Crested Becard	*Pachyramphus validus*	RT/SG/HPF	< 2000	C	R

VIREOS Vireonidae

Rufous-browed Peppershrike	*Cyclarhis gujanensis*	SG/HPF	< 2000	M/C	R
Brown-capped Vireo	*Vireo leucophrys*	SG/HPF	1800-2600	C	FC

Red-eyed Vireo	*Vireo olivaceus*	SG/HPF	< 2600	C	FC

JAYS Corvidae

White-collared Jay	*Cyanolyca viridicyana*	HMF/HPF	2200-3500	M/C	FC
Green (Inca) Jay	*Cyanocorax yncas*	SG/HPF	< 2400	M/C	FC

SWALLOWS AND MARTINS Hirundinidae

Blue-and-white Swallow	*Pygochelidon cyanoleuca*	R/SG/AA	< 3500	A	C
Brown-bellied Swallow	*Orochelidon murina*	HMS/AMS/AA	2500-4600	A	FC
Pale-footed Swallow	*Orochelidon flavipes*	EF/HMF	2000-3500	A	FC
Andean Swallow	*Orochelidon andecola*	PG/AMS	3500-4600	A	U
Southern Rough-winged Swallow	*Stelgidopteryx ruficollis*	R/SG/AA/	< 2000	A	R
Brown-chested Martin	*Progne tapera*	R/SG/AA/	3000	A	am V
Purple Martin	*Progne subis*	FM/AA/AMS	3200	A	bm R
White-rumped Swallow	*Tachycineta leucorrhoa*	AA/AMS/PB	3200-4000	A	am R
Bank Swallow	*Riparia riparia*	AA/SG/R/FM	< 3000	A	bm U
Barn Swallow	*Hirundo rustica*	SG/AA/FM	< 3300	A	bm FC
Cliff Swallow	*Petrochelidon pyrrhonota*	>	2000-3300	A	bm U

WRENS Troglodytidae

Gray-mantled Wren	*Odontorchilus branickii*	HPF	< 2000	C	U
House Wren	*Troglodytes aedon*	SG/AMS/HMS	1800 - 4600	T/U	C
Mountain Wren	*Troglodytes solstitialis*	EF/HMS/HPF	2000-3600	M	C
Sedge (Puna Wren) Wren	*Cistothorus (minimus) platensis*	EF/PG	3500-4600	U	FC
E Inca Wren	*Pheugopedius eisenmanni*	HMF/HPF/B	1800-3400	C	FC
Fulvous Wren	*Cinnycerthia fulva*	HMF/HPF	1800-3000	U	FC
Gray-breasted Wood-Wren	*Henicorhina leucophrys*	HMF/HPF	1800-2800	U	FC

DIPPERS Cinclidae

White-capped Dipper	*Cinclus leucocephalus*	R/S	1800-3200	T	FC

THRUSHES AND SOLITAIRES Turdidae

Andean Solitaire	*Myadestes ralloides*	SG/HMF/HPF	1800-2800	M/C	FC
Slaty-backed Nightingale-Thrush	*Catharus fuscater*	HPF	1800-2900	T/U	R
Swainson's Thrush	*Catharus ustulatus*	>	< 3500	T/U	bm FC
White-eared Solitaire	*Entomodestes leucotis*	HMF/HPF	1800-2900	M/C	U

Pale-eyed Thrush	*Turdus leucops*	HPF	1800-2000	M/C	U
Black-billed Thrush	*Turdus ignobilis*	RT/SG/HPF	< 2000	T/C	R
Slaty Thrush	*Turdus nigriceps*	HPF	< 2000	C	am R
Great Thrush	*Turdus fuscater*	EF/HMF/HMS	2500-4200	T/C	C
Chiguanco Thrush	*Turdus chiguanco*	SG/AMS/HMS	2400-4600	T/C	C
Glossy-black Thrush	*Turdus serranus*	HMF/HPF	1800-3000	C	FC

MOCKINGBIRDS — Mimidae

Brown-backed Mockingbird	*Mimus dorsalis*	AMS/AA	3000	T/U	V

PIPITS — Motacillidae

Short-billed Pipit	*Anthus furcatus*	PG	3500-4500	T	FC
Correndera Pipit	*Anthus correndera*	PG	3300-4550	T	U
Paramo Pipit	*Anthus bogotensis*	PG/AB	3000-4500	T	U

TANAGERS & ALLIES — Thraupidae

Magpie Tanager	*Cissopis leverianus*	SG/RT	< 2000	U/C	R
Slaty Tanager	*Creurgops dentatus*	HPF	< 2500	C	U
E	Black-capped (White-browed) Hemispingus	*Hemispingus (atropileus) auricularis*	EF/HMF/HPF	2600-3700	U/M	FC
E	Parodi's Hemispingus	*Hemispingus parodii*	EF/HMF/B	2600-3500	U	FC
Superciliaried Hemispingus	*Hemispingus superciliaris*	HMF/HPF/SG	2200-3450	C	FC
Oleaginous Hemispingus	*Hemispingus frontalis*	HMF/HPF/SG	1800-2600	U/M	U
Black-eared Hemispingus	*Hemispingus melanotis*	HMF/HPF/SG	1800-2200	U	U
Drab Hemispingus	*Hemispingus xanthophthalmus*	EF/HMF/HPF	2200-3500	M/C	U
Three-striped Hemispingus	*Hemispingus trifasciatus*	EF/HMF	3000-3700	C	FC
Gray-hooded Bush-Tanager	*Cnemoscopus rubrirostris*	HMF/HPF	2000-3000	C/M	FC
Rufous-chested Tanager	*Thlypopsis ornata*	AMS/HMS	1800-3300	U/C	R
Rust-and-yellow Tanager	*Thlypopsis ruficeps*	HMF/HPF/SG	1800-3700	U/C	C
White-lined Tanager	*Tachyphonus rufus*	SG/RT	< 2000	U/C	U
Silver-beaked Tanager	*Ramphocelus carbo*	SG/RT/	< 2000	U/C	U
Hooded Mountain-Tanager	*Buthraupis montana*	HMF	2300-3500	C	FC
Grass-green Tanager	*Chlorornis riefferii*	HMF/HPF	2000-3500	C	FC
Lacrimose Mountain-Tanager	*Anisognathus lacrymosus*	HMF/HPF/SG	1800-3500	M	R
Scarlet-bellied Mountain-Tanager	*Anisognathus igniventris*	EF/HMF/SG	2400-3600	U/C	C

Blue-winged Mountain-Tanager	*Anisognathus somptuosus*	HPF	< 2300	C/M	R
Buff-breasted Mountain-Tanager	*Dubusia taeniata*	EF/HMF	1900-3400	U/M	U
Chestnut-bellied Mountain-Tanager	*Dubusia castaneoventris*	EF/HMF/HPF	2000-3500	C	FC
Yellow-throated Tanager	*Iridosornis analis*	HPF	1800-2300	U/M	FC
Golden-collared Tanager	*Iridosornis jelskii*	EF/HMF	3000-3700	U/C	FC
E Yellow-scarfed Tanager	*Iridosornis reinhardti*	HMF/HPF	2100-3700	U/M	U
Fawn-breasted Tanager	*Pipraeidea melanonota*	PW/HMF/SG	< 3000	M/C	U
Blue-and-yellow Tanager	*Pipraeidea bonariensis*	SG/HMS/AMS	1800-4000	C	FC
Orange-eared Tanager	*Chlorochrysa calliparaea*	HPF	< 2200	C	FC
Blue-gray Tanager	*Thraupis episcopus*	SG/RT	< 2000	C	FC
Sayaca Tanager	*Thraupis sayaca*	SG/AA	3000	M/C	am V
Palm Tanager	*Thraupis palmarum*	SG/RT	< 2000	C	V
Blue-capped Tanager	*Thraupis cyanocephala*	HMF/HPF/SG	1800-3100	C	C
Golden-naped Tanager	*Tangara ruficervix*	HPF	1800-2400	C	FC
Silvery Tanager	*Tangara viridicollis*	HMF/HPF	1800-2700	C	FC
Blue-necked Tanager	*Tangara cyanicollis*	HPF	1800-2400	C	C
Blue-and-black Tanager	*Tangara vassorii*	HMF/SG	2400-3500	C	FC
Beryl-spangled Tanager	*Tangara nigroviridis*	HPF/SG	1800-2500	C	FC
Bay-headed Tanager	*Tangara gyrola*	HPF	< 2000	C	R
Saffron-crowned Tanager	*Tangara xanthocephala*	HPF/SG	1800-2400	C	FC
Flame-faced Tanager	*Tangara parzudakii*	HPF	1800-2600	C	R
Blue Dacnis	*Dacnis cayana*	HPF/SG	< 2000	C	U
Cinereous Conebill	*Conirostrum cinereum*	HMS/AMS/SG/RT	2500-4000	U/C	C
Blue-backed Conebill	*Conirostrum sitticolor*	EF/HMF	2300-3600	C	FC
Capped Conebill	*Conirostrum albifrons*	HPF/SG	1800-3000	C	FC
White-browed Conebill	*Conirostrum ferrugineiventre*	EF/HMF	2600-4100	M/C	FC
Giant Conebill	*Oreomanes fraseri*	PW	3500-4600	U/C	R
Tit-like Dacnis	*Xenodacnis parina*	PW	3200-4600	U/C	U
Moustached Flowerpiercer	*Diglossa mystacalis*	EF/HMF	2500-3700	U/C	FC
Black-throated Flowerpiercer	*Diglossa brunneiventris*	AMS/HMS/HMF/SG	2400-4300	U/C	C

Common Name	Scientific Name	Habitat	Elevation	Abundance	Status
Rusty Flowerpiercer	*Diglossa sittoides*	HMS/SG	1800-3500	M/C	FC
Bluish Flowerpiercer	*Diglossa caerulescens*	HMF/HPF	1800-3100	C	FC
Masked Flowerpiercer	*Diglossa cyanea*	EF/HMF/HPF	1800-3600	C	C
Buff-throated Saltator	*Saltator maximus*	SG/HPF	< 2000	M/C	R
Golden-billed Saltator	*Saltator aurantiirostris*	SG/AMS/HMS	2100-4000	U/C	C
Plushcap	*Catamblyrhynchus diadema*	HMF/HPF/B	2000-3500	U/M	FC
Peruvian Sierra-Finch	*Phrygilus punensis*	AA/HMS/AMS	2900-4600	T/U	C
Mourning Sierra-Finch	*Phrygilus fruticeti*	HMS/AMS	2300-4200	T/U	U
Plumbeous Sierra-Finch	*Phrygilus unicolor*	PW/PG	3000-4700	T	C
Ash-breasted Sierra-Finch	*Phrygilus plebejus*	PG/AMS	2400-4500	T	C
Band-tailed Sierra-Finch	*Phrygilus alaudinus*	PG/AMS/AA	3000-4000	T	U
Short-tailed Finch	*Idiopsar brachyurus*	PG	3900-4600	T	U
White-winged Diuca-Finch	*Diuca speculifera*	PG/AB	4000-4800	T	FC
Slaty Finch	*Haplospiza rustica*	HMF/HPF/SG/B	1800-3300	U/M	U/N
E Chestnut-breasted Mountain-Finch	*Poospiza caesar*	AMS/HMS/AA	3000-3850	U/M	FC
Bright-rumped Yellow-Finch	*Sicalis uropygialis*	PG	3300-4800	T	FC
Greenish Yellow-Finch	*Sicalis olivascens*	AMS/AA	2000-4200	T	FC
Grassland Yellow-Finch	*Sicalis luteola*	AA/AMS	3000-3900	T	R/U
Black-and-white Seedeater	*Sporophila luctuosa*	HMS/RT/SG/AA	< 3200	U	U/N
Yellow-bellied Seedeater	*Sporophila nigricollis*	HMS/SG/AA	< 2600	U	FC
Band-tailed Seedeater	*Catamenia analis*	AMS/HMS/SG	2500-4000	T/U	C
Plain-colored Seedeater	*Catamenia inornata*	PW/HMS/PG	2600-4400	T/U	U
Paramo Seedeater	*Catamenia homochroa*	EF/HMS/HMF	2350-3500	T/U	FC
Bananaquit	*Coereba flaveola*	HPF/SG	< 2400	M/C	U
Dull-colored Grassquit	*Tiaris obscura*	HMS/AA/AMS	< 2100	T/U	U
NEW WORLD SPARROWS	**Emberizidae**				
• Rufous-collared Sparrow	*Zonotrichia capensis*	AMS/SG/HMS/AA	< 3600	T/U	C
Yellow-browed Sparrow	*Ammodramus aurifrons*	SG/RT/AA	< 2000	T/U	R

	Common Name	Scientific Name	Habitat	Elevation	Abundance	Status
	Chestnut-capped Brush Finch	*Arremon brunneinucha*	HMF/HPF/SG	1800-2500	T/U	FC
	Gray-browed Brush Finch	*Arremon assimilis*	EF/HMF/HPF	2000-3400	T/U	FC
E	Tricolored Brush Finch	*Atlapetes tricolor*	HPF/SG	1800-3050	U/M	FC
E	Apurimac Brush Finch	*Atlapetes forbesi*	HMS/AA/PW	2700-4000	U/M	R
E	Cuzco Brush Finch	*Atlapetes canigenis*	HMF/HPF/SG	2450-3000	U/M	FC
	Common Chlorospingus	*Chlorospingus ophthalmicus*	HPF/SG	1800-2650	U/M	C
	Yellow-whiskered Chlorospingus	*Chlorospingus parvirostris*	HPF	1800-2750	U/C	FC

CARDINAL GROSBEAKS — Cardinalidae

	Common Name	Scientific Name	Habitat	Elevation	Abundance	Status
	(Highland) Hepatic Tanager	*Piranga flava lutea*	SG/HMF/HPF	1800-2700	C	FC
	Summer Tanager	*Piranga rubra*	SG/HMF/HPF	< 3000	C	bm U
	Scarlet Tanager	*Piranga olivacea*	SG/HMF/HPF	< 2000	M/C	bm R
	Golden Grosbeak	*Pheucticus chrysogaster*	AA/SG	< 3500	C	R
●	Black-backed Grosbeak	*Pheucticus aureoventris*	>	1800-3200	C	FC

NEW WORLD WARBLERS — Parulidae

	Common Name	Scientific Name	Habitat	Elevation	Abundance	Status
	Masked Yellowthroat	*Geothlypis aequinoctialis*	RT/SG	< 2000	U	R
	Cerulean Warbler	*Setophaga cerulea*	HPF/SG	< 2000	C/M	bm R
	Tropical Parula	*Setophaga pitiayumi*	HPF	< 2000	C	U
♥	Blackburnian Warbler	*Setophaga fusca*	HPF/SG	< 3000	C/M	bm FC
	Citrine Warbler	*Myiothlypis luteoviridis*	EF/HMF	2500-3700	U	FC
	Pale-legged Warbler	*Myiothlypis signatus*	HPF/HMF/SG	1800-2800	U	C
	Russet-crowned Warbler	*Myiothlypis coronatus*	HMF/HPFSG	1800-2900	U	FC
	Three-striped Warbler	*Basileutrus tristriatus*	HMF/HPF	< 2200	U	FC
	Canada Warbler	*Cardellina canadensis*	HPF/SG	< 2000	U/M	bm R
	Slate-throated Redstart (Whitestart)	*Myioborus miniatus*	HPF/SG	1800-2600	M/C	C
	Spectacled Redstart (Whitestart)	*Myioborus melanocephalus*	HPF/HMF	2000-3500	M/C	C

ORIOLES & BLACKBIRDS — Icteridae

	Common Name	Scientific Name	Habitat	Elevation	Abundance	Status
♣	Dusky-green Oropendola	*Psarocolius atrovirens*	HPF/SG	< 2600	C	FC
	Southern Mountain Cacique	*Cacicus chrysonotus*	HMF/HPF	1800-3450	C	FC
	Yellow-billed Cacique	*Amblycercus holosericeus*	HMF/HPF	2100-3300	U/M	U

Yellow-winged Blackbird	*Agelasticus thilius*	FM	3000- 4200	T/U	C
Bobolink	*Dolichonyx oryzivorus*	?	< 2500	T	bm R
White-browed Meadowlark	*Sturnella superciliaris*	AA/AMS	3000	T/U	am V

FINCHES   Fringillidae

Thick-billed Siskin	*Sporaga crassirostris*	PW	3600-4600	C	U
Hooded Siskin	*Sporaga magellanica*	AMS/HMS/SG/	2000-4200	C	C
Olivaceous Siskin	*Sporaga olivacea*	HPF/SG	< 2500	C	FC
Black Siskin	*Sporaga atrata*	PG/AMS	3500-4800	C	FC
Thick-billed Euphonia	*Euphonia laniirostris*	HPF/SG	< 2000	C	U
Orange-bellied Euphonia	*Euphonia xanthogaster*	HPF	< 2200	C	U
Blue-naped Chlorophonia	*Chlorophonia cyanea*	HMF	< 2100	C	R

HABITATS

R Rivers
SG Secondary Growth Forest
HMF Humid Montane Forest (2500-3400)
HPF Humid Pre-montane Forest (< 2500)
EF Elfin Forest
PW Polylepis Woodland
PB Puna Grassland
AB Andean Bogs
FL Freshwater Lakes and Ponds
HMS Semi Humid/Humid Montane Scrub
AMS Arid Montane Scrub
AA Agricultural Areas
RT Riparian Thickets
FM Freshwater Marshes
> More than four Habitats

MICRO-HABITATS

B Bamboo
T Treefalls
VT Viny Tangles
S Streamsides
TS Talus Slopes

FORAGING POSITION

T Terrestrial
U Understory
M Midstory
C Canopy
A Aerial
W Water

STATUS

C Common
FC Fairly Common
U Uncommon
V Vagrant
? Status Unclear
bm Boreal Migrant
am Austral Migrant

E PERUVIAN ENDEMIC

REFERENCES

Compiled by Allen Hale

Beolens, Bo, Michael Watkins, and Michael Grayson. *The Eponym Dictionary of Birds*. London: Bloomsbury, 2014.

Buckley, P.A., Mercedes S. Foster, Eugene S. Morton, Robert S. Ridgely, and Francine G. Buckley, Editors. *Neotropical Ornithology*. Ornithological Monographs No. 36. Washington: American Ornithologists' Union, 1985.

Chapman, Frank, M. *The Distribution of Bird Life in the Urubamba Valley of Peru: A Report on the Birds Collected by the Yale University-National Geographic Society's Expeditions*. Washington: Smithsonian Institution, 1921. United States National Museum Bulletin 117.

Chepstow-Lusty, A. J., M. R. Frogley, B. S. Bauer, M. J. Leng, K. P. Boessenkool, C. Carcaillet, A. A. Ali, and A. Gioda. Putting the Rise of the Inca Empire within a Climatic and Land Management Context. *Climate of the Past* 5 (2009): 375–388.

Clements, James A. and Noam Shany. *A Field Guide to the Birds of Peru*. Illustrated by Dana Gardner and Eustace Barnes. Temecula: Ibis Publishing Company, 2001.

del Hoyo, Josep, Andrew Elliott, Jordi Sargatal, and David Christie, Editors. *Handbook of the Birds of the World*. Volumes 1-16 and *Special Volume, New Species and Global Index*. Barcelona, Lynx, 1992-2013.

Dunn, Jon L. and Jonathan Alderfer. *National Geographic Field Guide to the Birds of North America*. Sixth Edition. Washington: National Geographic Society, 2011.

Fjeldså, Jon and Neils Krabbe. *Birds of the High Andes: A Manual to the Birds of the Temperate Zone of the Andes and Patagonia, South America*. Illustrated by Jon Fjeldså. Copenhagen: Zoological Museum, University of Copenhagen and Svendborg: Apollo Books, 1990.

Jobling, James A. *Helm Dictionary of Scientific Bird Names: From Aalge to Zusii*. London: Christopher Helm, 2011.

Lloyd, Huw. Abundance and Patterns of Rarity of *Polylepis* Birds in the Cordillera Vilcanota, Southern Perú: Implications for Habitat Management Strategies. *Bird Conservation International* 18, no. 2 (2008): 164-180.

Lloyd, Huw. Foraging Ecology of High Andean Insectivorous Birds in Remnant *Polylepis* Forest Patches. *The Wilson Journal of Ornithology* 120, no. 3 (2009): 531-544.

Lloyd, Huw and Stuart J. Marsden. Bird Community Variation across *Polylepis* Woodland Fragments and Matrix Habitats: Implications for Biodiversity Conservation within a High Andean Landscape. *Biodiversity and Conservation* 17, no. 11 (2008): 2645-2660.

Meyer de Schauensee, Rodophe. *A Guide to the Birds of South America*. Wynnewood: Livingston Publishing Company for the Academy of Natural Sciences of Philadelphia, 1970.

Parker, Theodore A., III, and John P. O'Neill. Notes on Little Known Birds of the Upper Urubamba Valley, Southern Peru. *The Auk* 97, no. 1 (1980): 167-176.

Parker, Theodore A., III, Susan Allen Parker, and Manuel A. Plenge. *An Annotated Checklist of the Peruvian Birds*. Vermillion: Buteo Books, 1982.

Peters, James L. *et al. Check-List of the Birds of the World.* Cambridge: Museum of Comparative Zoology, Harvard, 1934-1987. Volume I (Second edition) through Volume XVI (Comprehensive Index).

Peterson, Roger Tory. *Peterson Field Guide to Birds of Eastern and Central North America.* Sixth Edition. Boston: Houghton Mifflin Harcourt, 2010.

Remsen, J.V. Jr., Editor. *Studies in Neotropical Ornithology Honoring Ted Parker.* Ornithological Monographs No. 48. Washington: American Ornithologists' Union, 1997

Ridgely, Robert S. and Paul J. Greenfield. *The Birds of Ecuador.* Volume I: *Status, Distribution, and Taxonomy.* Volume II: *Field Guide.* Ithaca: Cornell University Press, 2001.

Ridgely, Robert S. and Guy Tudor. *The Birds of South America.* Volume I: *The Oscine Passerines,* 1989 and Volume II: *The Suboscine Passerines*, 1994. Austin: University of Texas Press.

Robbins, Mark B., David Geale, Barry Walker, Tristan J. Davis, Mariela Combe, Muir D. Easton, and Kyle P. Kennedy. Foothill Avifauna of the Upper Urubamba Valley. *Cotinga* 33 (2011): 41-52.

Robbins, Mark B. and Árpád S. Nyári. Canada to Tierra del Fuego: Species Limits and Historical Biogeography of the Sedge Wren (*Cistothorus platensis*). *The Wilson Journal of Ornithology* 126, no. 4 (2014): 649-662.

Schulenberg, Thomas S., Douglas F. Stotz, Daniel F. Lane, John P. O'Neill, and Theodore A. Parker III. *Birds of Peru: Revised and Updated Edition.* Principal illustrators: Dale Dyer, Daniel F. Lane, Lawrence B. McQueen, John P. O'Neill, and N. John Schmitt. Princeton: Princeton University Press, 2007.

Stephens, Lorain and Melvin A. Traylor, Jr. *Ornithological Gazetteer of Peru.* Cambridge: Museum of Comparative Zoology, Harvard University, 1983.

Taylor, Tom. *Aves: A Survey of the Literature of Neotropical Ornithology.* Baton Rouge: Louisiana State University Libraries, 2011.

Valqui, Thomas. *Where to Watch Birds in Peru.* Lima, 2004.

Vaurie, Charles. *An Ornithological Gazetteer of Peru (Based on Information Compiled by J.T. Zimmer).* New York: American Museum of Natural History, 1972. *American Museum Novitates* No. 2491.

Venero Gonzales, José Luis. Etnornitología y Guía de Aves del Humedal "Lucre-Huacarpay." *Editorial Moderna* (2008).

Walker, Barry. *The Birds of Machu Picchu and the Cusco Region: Field Guide.* Illustrated by Jon Fjeldså. Lima: Nuevas Imágenes S.A., 2005.

Walker, Barry, Huw Lloyd, and Gerard Cheshire. *Peruvian Wildlife: A Visitor's Guide to the Central Andes.* Bradt Travel Guides, 2007.

Zimmer, John T. *Birds of the Marshall Field Peruvian Expedition, 1922-1923.* Chicago Field Museum of Natural History, 1930. Publication 282.

Zimmer, John T. *Studies of Peruvian Birds: Nos. 1-66.* New York: American Museum of Natural History, *American Museum Novitates*, 1931-1935.

INDEX

Accipiter collaris 48
Accipiter ventralis 48
ACCIPITRIDAE 48
Actitis macularia 62
Adelomyia melanogenys 84
Aeronautes andecolus 80
Aeronautes montivagus 80
Agelasticus thilius 192
Aglaeactis cupripennis 88
Aglaeactis castelnaudii 88
Aglaiocercus kingi 84
Agriornis albicauda 134
Agriornis micropterus 134
Agriornis montana 134
Alder Flycatcher 128
Amazilia chionogaster 92
Amazilia viridicauda 92
Amazona mercenaria 100
Amblycercus holosericeus 192
American Golden Plover 58
American Kestrel 52
AMERICAN VULTURES 46
Amethyst-throated Sunangel 84
Ammodramus aurifrons 182
Ampelion rubrocristata 142
Ampelion rufaxilla 144
Anabacerthia striaticollis 110
Anairetes alpinus 122
Anairetes flavirostris 122
Anairetes parulus 122
Anas bahamensis 36
Anas cyanoptera 36
Anas discors 36
Anas flavirostris 36
Anas georgica 36
Anas platalea 36
Anas puna 36
ANATIDAE 34
Andean Avocet 60
Andean Cock-of-the-rock 144
Andean Condor 46
Andean Coot 56
Andean Duck 34
Andean Flicker 98
Andean Goose 34
Andean Guan 38
Andean Gull 66
Andean Hillstar 84
Andean Ibis 46
Andean Lapwing 58
Andean Motmot 94
Andean Negrito 130
Andean Parakeet 100

Andean Snipe 60
Andean Solitaire 154
Andean Swallow 148
Andean Swift 80
Andean Tinamou 32
Andean Tit-Spinetail 114
Andigena hypoglauca 96
Anisognathus igniventris 164
Anisognathus lacrymosus 164
Anisognathus somptuosus 166
ANTBIRDS 102
Anthus bogotensis 158
Anthus correndera 158
Anthus furcatus 158
ANTPITTAS 102
ANTTHRUSHES 106
Aplomado Falcon 54
APODIDAE 80
Apurimac Brushfinch 184
Ardea alba 44
Ardea cocoi 44
ARDEIDAE 42
Arremon assimilis 182
Arremon brunneinucha 182
Ash-breasted Sierra-Finch 176
Ash-breasted Tit-Tyrant 122
Ashy-headed Tyrannulet 118
Asthenes helleri 116
Asthenes humilis 116
Asthenes maculicauda 114
Asthenes modesta 116
Asthenes ottonis 116
Asthenes urubambensis 114
Asthenes virgata 114
Asthenes wyatti 116
Athene cunicularia 76
Atlapetes canigenis 184
Atlapetes forbesi 184
Atlapetes tricolor 184
Attagis gayi 66
*Aulacorhynchus (prasinus)
   atrogularis* 96
*Aulacorhynchus
   coeruleicinctus* 96
Azara's Spinetail 116
Baird's Sandpiper 64
Bananaquit 180
Band-bellied Owl 74
Band-tailed Fruiteater 142
Band-tailed Pigeon 68
Band-tailed Seedeater 180
Band-tailed Sierra-Finch 178
Band-winged Nightjar 78

Bank Swallow 148
Bar-bellied Woodpecker 98
BARBETS 96
Bare-faced Ground-Dove 72
Barn Owl 74
BARN OWLS 74
Barn Swallow 148
Barred Antthrush 106
Barred Becard 144
Barred Fruiteater 142
Barred Parakeet 100
Bartramia longicauda 62
Baryphthengus martii 94
Basileuterus tristriatus 190
Bay-headed Tanager 170
Bearded Mountaineer 86
BECARDS 144
Beryl-spangled Tanager 170
Black and Chestnut Eagle 48
Black and White Seedeater 180
Black Phoebe 130
Black Siskin 194
Black Skimmer 68
Black Tern 68
Black-backed Grosbeak 186
Black-bellied (Grey) Plover 58
Black-bellied Whistling Duck 34
Black-billed Shrike-Tyrant 134
Black-billed Thrush 154
Blackburnian Warbler 188
Black-capped (White-browed)
   Hemispingus 160
Black-capped Tyrannulet 118
Black-chested Buzzard-Eagle 52
Black-crowned Night-heron 42
Black-eared Hemispingus 162
Black-streaked Puffbird 96
Black-tailed Trainbearer 84
Black-throated (Emerald)
   Toucanet 96
Black-throated Flowerpiercer 174
Black-throated Tody-Tyrant 126
Black-winged Ground-Dove 72
Blue Dacnis 172
Blue-and-black Tanager 170
Blue-and-white Swallow 148
Blue-and-yellow Tanager 168
Blue-backed Conebill 172
Blue-banded Toucanet 96
Blue-capped Tanager 168
Blue-gray Tanager 168
Blue-mantled Thornbill 86
Blue-naped Chlorophonia 196

237

Blue-necked Tanager 170
Blue-winged Mountain-
    Tanager 166
Blue-winged Teal 36
Bluish Flowerpiercer 174
Bobolink 192
Boissonneaua matthewsii 90
Bolborhynchus lineola 100
Bolborhynchus orbygnesius 100
Bolivian Tyrannulet 124
Booted Racket-tail 92
Bran-colored Flycatcher 126
Bright-rumped Yellow-Finch 178
Broad-winged Hawk 52
Bronzy Inca 90
Brown Tinamou 32
Brown-backed Chat-Tyrant 138
Brown-backed Mockingbird 158
Brown-bellied Swallow 148
Brown-capped Vireo 146
Brown-chested Martin 146
*Bubo (magellanicus)
    virginianus* 74
Bubulcus ibis 44
BUCCONIDAE 96
Buff-breasted Earthcreeper 110
Buff-breasted Mountain-
    Tanager 166
Buff-breasted Sandpiper 64
Buff-tailed Sicklebill 82
Buff-thighed Puffleg 88
Buff-throated Saltator 176
Burrowing Owl 76
Buteo albigula 52
Buteo platypterus 52
Buteogallus solitarius 50
Buthraupis montana 164
Butorides striatus 44
Cacicus chrysonotus 192
Caldiris minutilla 64
Calidris alba 64
Calidris bairdii 64
Calidris fuscicollis 64
Calidris himantopus 64
Calidris melanotos 64
Calidris pusilla 64
Campephilus haematogaster 98
Canada Warbler 190
CAPITONIDAE 96
Capped Conebill 172
CAPRIMULGIDAE 78
Cardellina canadensis 190
CARDINAL GROSBEAKS 186
CARDINALIDAE 186
Casiornis rufa 140
Catamblyrhynchus diadema 176

Catamenia analis 180
Catamenia homochroa 180
Catamenia inornata 180
Cathartes aura 46
CATHARTIDAE 46
Catharus fuscater 154
Catharus ustulatus 154
Cattle Egret 44
Cerulean Warbler 188
Chaetocercus mulsant 92
Chaetura pelagica 80
Chalcostigma olivaceum 86
Chalcostigma ruficeps 86
Chalcostigma stanleyi 86
Chamaepetes goudotii 38
Chamaeza mollissima 106
CHARADRIIDAE 58
Charadrius alticola 58
Charadrius collaris 58
Charadrius semipalmatus 58
Chestnut-bellied Mountain-
    Tanager 166
Chestnut-breasted Coronet 90
Chestnut-breasted Mountain-
    Finch 180
Chestnut-capped Brushfinch 182
Chestnut-collared Swift 80
Chestnut-crested Cotinga 144
Chiguanco Thrush 156
Childonias niger 68
Chilean Flamingo 40
Chimney Swift 80
Chlorochrysa calliparaea 168
Chlorophonia cyanea 196
Chlorornis riefferii 164
Chlorospingus ophthalmicus 184
Chlorospingus parvirostris 184
Chordeilis nacunda 78
Chroicocephalus serranus 66
Ciccaba albitarsus 76
CICONIDAE 42
CINCLIDAE 152
Cinclodes albiventris 112
Cinclodes aricomae 112
Cinclodes atacamensis 112
Cinclus leucocephalus 152
Cinereous Conebill 172
Cinereous Ground-Tyrant 132
Cinereous Harrier 50
Cinnamon Flycatcher 128
Cinnamon Teal 36
Cinnycerthia fulva 152
Circus cinereus 50
Cissopis leveriana 160
Cistothorus platensis 150
Citrine Warbler 188

Claravis mondetoura 70
Cliff Flycatcher 128
Cliff Swallow 148
Cnemarchus erythropygius 136
Cnemoscopus rubrirostris 162
Coccyzus americanus 72
Coccyzus melacoryphus 72
Cocoi Heron 44
Coeligena coeligena 90
Coeligena torquata 90
Coeligena violifer 90
Coereba flaveola 180
Colaptes rupicola 98
Colibri coruscans 82
Colibri thalassinus 82
Collared Inca 90
Collared Plover 58
Columba livia 68
COLUMBIDAE 68
Common Chlorospingus 184
Common Gallinule 56
Common Miner 108
Common Tern 68
Common Tody-Flycatcher 126
Conirostrum albifrons 172
Conirostrum cinereum 172
Conirostrum ferrugineiventre 172
Conirostrum sitticolor 172
Conopias cinchoneti 140
Contopus cooperi 128
Contopus fumigatus 128
Contopus sordidulus 128
Cordilleran Canastero 116
CORMORANTS 42
Correndera Pipit 158
CORVIDAE 146
COTINGAS 142
COTINGIDAE 142
CRACIDAE 38
Cranioleuca albicapilla 118
Cranioleuca marcapatae 118
Cream-winged Cinclodes 112
Creamy-crested Spinetail 118
Crested Becard 144
Crested Duck 34
Crested Quetzal 94
Creurgops dentate 160
Crimson-bellied Woodpecker 98
Crimson-mantled Woodpecker 98
Crotophaga ani 74
Crotophaga major 72
Crowned Chat-Tyrant 136
Crowned Slaty Flycatcher 140
Crypturellus obsoletus 32
CUCKOOS AND ANIS 72
CUCULIDAE 72

Cuzco Brushfinch 184
Cyanocorax yncas 146
Cyanolyca viridicayanus 146
Cyclarhis gujanensis 144
d'Orbigny's Chat-Tyrant 138
Dacnis cayana 172
Dark-billed Cuckoo 72
Dendrocygna autumnalis 34
Diademed Sandpiper-Plover 58
Diademed Tapaculo 106
Diglossa brunneiventris 174
Diglossa caerulescens 174
Diglossa cyanea 176
Diglossa mystacalis 174
Diglossa sittoides 174
DIPPERS 152
Diuca speculifera 178
Dolichonyx oryzivorus 192
Doryfera ludovicae 82
Drab Hemispingus 162
Drymophila striaticeps 102
Drymotoxeres pucherani 108
Dubusia castaneoventris 166
Dubusia taeniata 166
DUCKS AND GEESE 34
Dull-colored Grassquit 180
Dusky-capped Flycatcher 142
Dusky-green Oropendola 192
Eared Dove 70
Eastern Kingbird 140
Egretta caerulea 44
Egretta thula 44
Elaenia albiceps 120
Elaenia flavogaster 120
Elaenia obscura 120
Elaenia pallatangae 120
Elaenia parvirostris 120
Elanoides forficatus 48
EMBERIZIDAE 182
Empidonax alnorum 128
Empidonomus aurantioatrocristatus 140
Ensifera ensifera 90
Entomodestes leucotis 154
Eriocnemis luciani 88
Eubucco versicolor 96
Eupetomena macroura 92
Euphonia laniirostris 196
Euphonia xanthogaster 196
Eurypyga helia 56
EURYPYGIDAE 56
Eutoxeres condamini 82
Falco deiroleucus 54
Falco femoralis 54
Falco peregrinus 54
Falco sparverius 52

FALCONIDAE (including MOUNTAIN CARACARA) 52
FALCONS 52
Fasciated Tiger-Heron 44
Fawn-breasted Brilliant 92
Fawn-breasted Tanager 168
FINCHES 194
Flame-faced Tanager 172
FLAMINGOES 40
Fork tailed Flytcacher 140
FORMICARIIDAE 106
Formicarius rufipectus 106
Franklin's Gull 66
FRINGILLIDAE 194
Fulica ardesiaca 56
Fulica gigantea 56
Fulvous Wren 152
FURNARIIDAE 108
Gallinago andina 60
Gallinago imperialis 60
Gallinago jamesoni 60
Gallinula galeata 56
Geositta cunicularia 108
Geositta tenuirostris 108
Geothlypis aequinoctialis 188
Geotrygon montana 70
Geranoaetus melanoleucus 52
Geranoaetus polyosoma 50
Giant Conebill 174
Giant Coot 56
Giant Hummingbird 92
Glaucidium bolivianum 76
Glaucidium peruanum 76
Glossy-black Thrush 156
Golden Grosbeak 186
Golden-billed Saltator 176
Golden-browed Chat-Tyrant 136
Golden-collared Tanager 166
Golden-crowned Flycatcher 140
Golden-headed Quetzal 94
Golden-naped Tanager 170
Golden-olive Woodpecker 98
Golden-plumed Parakeet 100
Grallaria andicolus 104
Grallaria erythroleuca 104
Grallaria guatemalensis 102
Grallaria rufula occabambae 104
Grallaria squamigera 102
Grallaricula ferrugineipectus 104
GRALLARIDAE 102
Grass-green Tanager 164
Grassland Yellow-Finch 178
Gray-bellied Shrike-Tyrant 134
Gray-breasted Seedsnipe 66
Gray-breasted Wood-Wren 152
Gray-browed Brushfinch 182

Gray-hooded Bush-Tanager 162
Gray-mantled Wren 150
Great (Lesser) Horned Owl 74
Great Egret 44
Great Kiskadee 140
Great Sapphirewing 90
Great Thrush 156
Greater Ani 72
Greater Scythebill 108
Greater Yellowlegs 62
GREBES 40
Green and White Hummingbird 92
Green Hermit 82
Green Jay 146
Green Violetear 82
Green-fronted Lancebill 82
Greenish Yellow-Finch 178
Green-tailed Trainbearer 84
Grey-breasted Mountain Toucan 96
GUANS, CURRASOWS AND ALLIES 38
GULLS AND TERNS 66
Haplophaedia assimilis 88
Haplospiza rustica 178
HAWKS 48
Heliangelus amethysticollis 84
Heliodoxa rubinoides 92
Hemispingus atropileus auricularis 160
Hemispingus frontalis 162
Hemispingus melanotis 162
Hemispingus parodii 160
Hemispingus superciliaris 160
Hemispingus trifasciatus 162
Hemispingus xanthophthalmus 162
Hemitriccus granadensis 126
Henicorhina leucophrys 152
Hepatic Tanager 186
HERONS 42
Highland Elaenia 120
Himantopus melanurus 60
Hirundinea ferruginea 128
HIRUNDINIDAE 146
Hirundo rustica 148
Hooded Mountain-Tanager 164
Hooded Siskin 194
Hooded Tinamou 32
House Wren 150
Hudsonian Godwit 62
HUMMINGBIRDS 82
Hymenops perspicillaus 132
IBIS 42
ICTERIDAE 192
Idiopsar brachyurus 178

Imperial Snipe 60
Inca Flycatcher 126
Inca Wren 152
Iridosornis analis 166
Iridosornis jelskii 166
Iridosornis reinhardti 166
Jabiru mycteria 42
Jabiru 42
JAYS 146
Junín Canastero 114
Knipolegus aterrimus 130
Knipolegus cabanisi 130
Knipolegus poecilurus 130
Lacrimose Mountain-Tanager 164
Lafresnaya lafresnayi 90
Large-billed Tern 66
LARIDAE 66
Laughing Gull 66
Least Sandpiper 64
Legatus leucophaius 138
Lemon-browed Flycatcher 140
Lepidocolaptes lacrymiger 108
Leptasthenura andicola 114
Leptasthenura xenothorax 114
Leptasthenura yanacensis 114
Leptopogon superciliaris 126
Leptopogon taczanowskii 126
Leptosittaca branickii 100
Leptotila verreauxi 70
Lesbia nuna 84
Lesbia victoriae 84
Lesser Yellowlegs 62
Lessonia oreas 130
Leucophaeus atricilla 66
Leucophaeus pipixcan 66
Limosa haemastica 62
Line-fronted Canastero 114
Little Blue Heron 44
Lochmias nematura 110
Long-tailed Sylph 84
Lophonetta specularioides 34
Lophotriccus pileatus 126
Lurocalis rufiventris 78
Lyre-tailed Nightjar 78
Magpie Tanager 168
Malacoptila fulvogularis 96
Many-colored Rush-Tyrant 124
Marble-faced Bristle-Tyrant 124
Marcapata Spinetail 118
Margarornis squamiger 112
Maroon-chested Ground-Dove 70
Masked Flowerpiercer 176
Masked Fruiteater 142
Masked Trogon 94
Masked Yellowthroat 188

Mecocerculus leucophrys 122
Mecocerculus poecilocercus 122
Mecocerculus stictopterus 122
Megascops albogularis 74
Merganetta armata 34
Metallura aeneocauda 88
Metallura tyrianthina 86
Metriopelia cecilae 72
Metriopelia melanoptera 72
Micropygia schomburgkii 54
MIMIDAE 158
Mimus dorsalis 158
Mionectes striaticollis 124
Mitred Parakeet 100
Mitrephanes olivaceus 130
MOCKINGBIRDS 158
MOMOTIDAE 94
MOMOTS 94
Momotus aequatorialis 94
Montane Foliage-gleaner 110
Montane Solitary Eagle 50
Montane Woodcreeper 108
MOTACILLIDAE 158
Mottle-cheeked Tyrannulet 124
Mountain Caracara 52
Mountain Velvetbreast 90
Mountain Wren 150
Mourning Sierra-Finch 176
Moustached Flowerpiercer 174
Muscisaxicola albifrons 132
Muscisaxicola albilora 134
Muscisaxicola cinerea 132
Muscisaxicola flavinucha 132
Muscisaxicola fluviatilis 132
Muscisaxicola griseus 132
Muscisaxicola juninensis 132
Muscisaxicola maculirostris 132
Muscisaxicola rufivertex 134
Mustelirallus erythrops 54
Myadestes ralloides 154
Mycteria americana 42
Myiarchus cephalotes 142
Myiarchus tuberculifer 142
Myioborus melanocephalus 190
Myioborus miniatus 190
Myiodynastes chrysocephalus 140
Myiodynastes maculatus 140
Myiophobus fasciatus 126
Myiotheretes fumigatus 134
Myiotheretes fuscorufus 136
Myiotheretes striaticollis 134
Myiothlypis coronata 190
Myiothlypis luteoviridis 188
Myiothlypis signata 188
Myiozetetes similis 138

Nacunda Nighthawk 78
Neotropic Cormorant 42
Nephelomyias ochraceiventris 128
NEW WORLD QUAIL 38
NEW WORLD SPARROWS 182
NEW WORLD WARBLERS 188
NIGHTHAWKS AND NIGHTJARS 78
Nothocercus nigrocapillus 32
Nothoprocta ornata 32
Nothoprocta pentlandii 32
Nothoproctao taczanowskii 32
Numenius phaeopus 62
Nyctanassa violacea 42
Nycticorax nycticorax 42
Ocellated Crake 54
Ocellated Piculet 98
Ochraceous-breasted Flycatcher 128
Ochre-naped Ground-Tyrant 132
Ochthoeca cinnamomeiventris thoracia 136
Ochthoeca frontalis 136
Ochthoeca fumicolor 138
Ochthoeca leucophrys 138
Ochthoeca oenanthoides 138
Ochthoeca pulchella 136
Ochthoeca rufipectoralis 138
Ocreatus underwoodii 92
Odontophorus balliviani 38
Odontophorus speciosus 38
Odontorchilus branickii 150
OILBIRD 76
Oilbird 76
Oleaginous Hemispingus 162
Olivaceous Siskin 194
Olivaceous Thornbill 86
Olive Flycatcher 130
Olive-sided Flycatcher 128
Orange-bellied Euphonia 196
Orange-breasted Falcon 54
Orange-eared Tanager 168
Oreomanes fraseri 174
Oreonympha nobilis 86
Oreotrochilus estella 84
Oressochen melanopterus 34
ORIOLES AND BLACKBIRDS 192
Ornate Tinamou 32
Orochelidon andecola 148
Orochelidon flavipes 148
Orochelidon murina 148
OSPREY 48
Osprey 48
OVENBIRDS (including WOODCREEPERS) 108
Oxyura ferruginea 34

Pachyramphus validus 144
Pachyramphus versicolor 144
Paint-billed Crake 54
Pale-edged Flycatcher 142
Pale-eyed Thrush 154
Pale-footed Swallow 148
Pale-legged Warbler 188
Palm Tanager 168
Pandion haliaetus 48
PANDIONIDAE 48
Parabuteo leucorrhous 50
Paramo Pipit 158
Paramo Seedeater 180
Pardirallus maculatus 54
Pardirallus sanguinolentus 54
Parodi's Hemispingus 160
PARROTS 100
PARULIDAE 188
Patagioenas fasciata 68
Patagioenas maculosa 68
Patagioenas plumbea 70
Patagona gigas 92
Pearled Treerunner 112
Pectoral Sandpiper 64
Penelope montagnii 38
Peregrine Falcon 54
Peruvian Pygmy-Owl 76
Peruvian Sierra-Finch 176
Petrochelidon pyrrhonota 148
Phacellodomus striaticeps 114
Phaethornis guy 82
Phaetusa simplex 66
PHALACROCORACIDAE 42
Phalacrocorax olivaceus 42
Phalaropus tricolor 64
Phalcoboenus megalopterus 52
Pharomachrus antisianus 94
Pharomachrus auriceps 94
PHASIANIDAE 38
Phegornis mitchellii 58
Pheucticus aureoventris 186
Pheucticus chrysogaster 186
Pheugopedius eisenmanni 152
Phleocryptes melanops 110
PHOENICOPTERIDAE 40
Phoenicopterus chilensis 40
Phrygilus alaudinus 178
Phrygilus fruticeti 176
Phrygilus plebejus 176
Phrygilus punensis 176
Phrygilus unicolor 176
Phyllomyias cinereiceps 118
Phyllomyias nigrocapillus 118
Phyllomyias plumbeiceps 120
Phyllomyias sclateri 118
Phyllomyias uropygialis 120

Phylloscartes ophthalmicus 124
Phylloscartes ventralis 124
Piaya cayana 72
PICIDAE 98
Piculus rivolii 98
Piculus rubiginosus 98
Picumnus dorbygnianus 98
PIGEONS AND DOVES 68
Pionus tumultuosus 100
PIPITS 158
Pipraeidea bonariensis 168
Pipraeidea melanonota 168
Pipreola arcuata 142
Pipreola intermedia 142
Pipreola pulchra 142
Piranga flava 186
Piranga olivacea 186
Piranga rubra 186
Piratic Flycatcher 138
Pitangus sulphuratus 140
Plain-breasted Hawk 48
Plain-colored Seedeater 180
Plegadis ridgwayi 46
PLOVERS 58
Plumbeous Black-Tyrant 130
Plumbeous Pigeon 70
Plumbeous Rail 54
Plumbeous Sierra-Finch 176
Plumbeous-crowned
 Tyrannulet 120
Plushcap 176
Pluvialis dominica 58
Pluvialis squatarola 58
Podiceps occipitalis 40
PODICIPEDIDAE 40
Polioxolmis rufipennis 136
Poospiza caesar 180
Porphyrio martinica 56
Premnoplex brunnescens 112
Premnornis guttuliger 110
Progne subis 146
Progne tapera 146
Psarocolius atrovirens 192
Pseudocolaptes
 boissonneautii 110
Pseudocolopteryx
 acutipennis 124
Pseudotriccus ruficeps 124
Psittacara mitrata 100
PSITTACIDAE 100
Pterophanes cyanopterus 90
PUFFBIRDS 96
Pulsatrix melanota 74
Puna Ground-Tyrant 132
Puna Ibis 46
Puna Plover 58

Puna Snipe 60
Puna Tapaculo 106
Puna Teal 36
Puna Thistletail 116
Purple Gallinule 56
Purple Martin 146
Purple-backed Thornbill 86
Pygochelidon cyanoleuca 148
Pyrocephalus rubinus 130
Pyrrhomyias cinnamomea 128
RAILS 54
RALLIDAE 54
RAMPHASTIDAE 96
Ramphocelus carbo 164
Ramphomicron microrhynchum 86
Recurvirostra andina 60
RECURVIROSTRIDAE 60
Red and White Antpitta 104
Red Shoveler 36
Red-crested Cotinga 142
Red-eyed Vireo 146
Red-rumped Bush-Tyrant 136
RHINOCRYPTIDAE 106
RHYNOCHOPIDAE 68
Riparia riparia 148
Roadside Hawk 50
Rock Pigeon 68
Rollandia rolland 40
Royal Cinclodes 112
Ruddy Quail-Dove 70
Rufous Antpitta 104
Rufous Casiornis 140
Rufous Motmot 94
Rufous-backed (Peruvian)
 Treehunter 112
Rufous-banded Owl 76
Rufous-bellied Bush-Tyrant 136
Rufous-bellied Nighthawk 78
Rufous-bellied Seedsnipe 66
Rufous-breasted Antthrush 106
Rufous-breasted Chat-Tyrant 138
Rufous-breasted Wood-Quail 38
Rufous-browed Peppershrike 144
Rufous-capped Thornbill 86
Rufous-chested Tanager 162
Rufous-collared Sparrow 182
Rufous-headed Pygmy-Tyrant 124
Rufous-naped Ground-Tyrant 134
Rufous-tailed Tyrant 130
Rufous-webbed Bush-Tyrant 136
Rupicola peruviana 144
Rupornis magnirostris 50
Russet-crowned Warbler 190
Rust-and-yellow Tanager 162
Rusty Flowerpiercer 174
Rusty-breasted Antpitta 104

Rusty-fronted Canastero 116
Rusty-winged Barbtail 110
Rynchops niger 68
Saffron-crowned Tanager 172
Saltator aurantiirostris 176
Saltator maximus 176
Sanderling 64
SANDPIPERS AND SNIPES 60
Sapphire-vented Puffleg 88
Sayaca Tanager 168
Sayornis nigricans 130
Scale-crested Pygmy-Tyrant 126
Scaled Antpitta 102
Scaled Metaltail 88
Scaly-naped Amazon 100
Scarlet Tanager 186
Scarlet-bellied Mountain-
    Tanager 164
Sclater's Tyrannulet 118
SCOLOPACIDAE 60
Scribble-tailed Canastero 114
Scytalopus parvirostris 106
Scytalopus schulenbergi 106
Scytalopus simonsi 106
Scytalopus urubambae 106
Sedge Wren 150
SEEDSNIPE 66
Semicollared Hawk 48
Semipalmated Plover 58
Semipalmated Sandpiper 64
Serpophaga cinerea 124
Setophaga cerulea 188
Setophaga fusca 188
Setophaga pitiayumi 188
Sharp-tailed Streamcreeper 110
Shining Sunbeam 88
Short-billed Pipit 158
Short-tailed Finch 178
Sicalis luteola 178
Sicalis olivascens 178
Sicalis uropygialis 178
Sickle-winged Guan 38
Sierran Elaenia 120
Silver-beaked Tanager 164
Silvery Grebe 40
Silvery Tanager 170
SKIMMERS 68
Slate-throated Redstart
    (Whitestart) 190
Slaty Finch 178
Slaty Tanager 160
Slaty Thrush 156
Slaty-backed Chat-Tyrant 136
Slaty-backed Nightingale
    Thrush 154
Slaty-capped Flycatcher 126

Slender-billed Miner 108
Small-billed Elaenia 120
Smoke-colored Peewee 128
Smoky Bush-Tyrant 134
Smooth-billed Ani 74
Snowy Egret 44
Social Flycatcher 138
Solitary Sandpiper 62
Southern Mountain Cacique 192
Southern Rough-winged
    Swallow 148
Sparkling Violetear 82
Speckled Hummingbird 84
Speckle-faced Parrot 100
Spectacled Redstart
    (Whitestart) 190
Spectacled Tyrant 132
Spizaetus isidori 48
Sporaga atrata 194
Sporaga crassirostris 194
Sporaga magellanica 194
Sporaga olivacea 194
Sporophila luctuosa 180
Sporophila nigricollis 180
Spot-billed Ground-Tyrant 132
Spotted Barbtail 112
Spotted Rail 54
Spotted Sandpiper 62
Spot-winged Pigeon 68
Squirrel Cuckoo 72
Steatornis caripensis 76
STEATORNITHIDAE 76
Stelgidopteryx ruficollis 148
Sterna hirundo 68
Stilt Sandpiper 64
STILTS 60
STORKS 42
Streak-backed Canastero 116
Streaked Flycatcher 140
Streaked Tuftedcheek 110
Streaked Xenops 110
Streak-fronted Thornbird 114
Streak-headed Antbird 102
Streak-necked Flycatcher 124
Streak-throated Bush-Tyrant 134
Streak-throated Canastero 116
Streptoprocne rutilus 80
Streptoprocne zonaris 80
Striated Heron 44
STRIGIDAE 74
Striped Treehunter 112
Striped-faced Wood-Quail 38
Stripe-headed Antpitta 104
Strong-billed Woodcreeper 108
Sturnella superciliaris 192
Subtropical Doradito 124

Summer Tanager 186
Sunbittern 56
SUNBITTERNS 56
Superciliaried Hemispingus 160
Swainson's Thrush 154
SWALLOWS AND MARTINS 146
Swallow-tailed Hummingbird 92
Swallow-tailed Kite 48
Swallow-tailed Nightjar 78
SWIFTS 80
Sword-billed Hummingbird 90
Synallaxis azarae 116
Systellura longirostris 78
Tachuris rubrigastra 124
Tachycineta leucorrhoa 148
Tachyphonus rufus 164
Taczanowski's Ground-Tyrant 132
Taczanowski's Tinamou 32
TANAGERS AND ALLIES 160
Tangara cyanicollis 170
Tangara gyrola 170
Tangara nigroviridis 170
Tangara parzudakii 172
Tangara ruficervix 170
Tangara vassorii 170
Tangara viridicollis 170
Tangara xanthocephala 172
TAPACULOS 106
Tawny Tit-Spinetail 114
Tawny-rumped Tyrannulet 120
THAMNOPHILIDAE 102
Thamnophilus caerulescens 102
Theristicus branickii 46
Thick-billed Euphonia 196
Thick-billed Siskin 194
THINOCORIDAE 66
Thinocorus orbignyianus 66
Thlypopsis ornata 162
Thlypopsis ruficeps 162
THRAUPIDAE 160
Thraupis cyanocephala 168
Thraupis episcopus 168
Thraupis palmarum 168
Thraupis sayaca 168
Three-striped Hemispingus 162
Three-striped Warbler 190
THRESKIORNITHIDAE 46
Thripadectes holostictus 112
Thripadectes scrutator 112
THRUSHES 154
Tiaris obscura 180
Tigrisoma fasciatum 44
TINAMIDAE 32
TINAMOUS 32
Tit-like Dacnis 174
TITYRIDAE 144

Todirostrum cinereum 126
Tolmomyias sulphurescens 126
Torrent Duck 34
Torrent Tyrannulet 124
TOUCANS 96
Tricolored Brushfinch 184
Trilling Tapaculo 106
Tringa flavipes 62
Tringa melanoleuca 62
Tringa solitaria 62
TROCHILIDAE 82
Troglodytes aedon 150
Troglodytes solstitialis 150
TROGLODYTIDAE 150
Trogon personatus 94
TROGONIDAE 94
TROGONS AND QUETZALS 94
Tropical Kingbird 140
Tropical Parula 188
Tryngites subruficollis 64
Tufted Tit-Tyrant 122
TURDIDAE 154
Turdus chiguanco 156
Turdus fuscater 156
Turdus ignobilis 154
Turdus leucops 154
Turdus nigriceps 156
Turdus serranus 156
Turkey Vulture 46
TYPICAL OWLS 74
TYRANNIDAE 118
Tyrannus melancholicus 140
Tyrannus tyrannus 140
TYRANT FLYCATCHERS 118
Tyranus savana 140
Tyrian Metaltail 86
Tyto alba 74
TYTONIDAE 74
Undulated Antpitta 102
Unstreaked Tit-Tyrant 122
Upland Sandpiper 62

Upucerthia validirostris 110
Uromyias agraphia 122
Uropsalis lyra 78
Uropsalis segmentata 78
Vanellus resplendens 58
Variable Antshrike 102
Variable Hawk 50
Veniliornis nigriceps 98
Vermillion Flycatcher 130
Versicolored Barbet 96
Vilcabamba Tapaculo 106
Violet-throated Starfrontlet 90
Vireo leucophrys 146
Vireo olivaceous 146
VIREONIDAE 144
VIREOS 144
Vulture gryphus 46
Western Wood Pewee 128
Whimbrel 62
White-backed Stilt 60
White-banded Tyrannulet 122
White-bellied Hummingbird 92
White-bellied Woodstar 92
White-browed Chat-Tyrant 138
White-browed Conebill 172
White-browed Ground-Tyrant 134
White-browed Meadowlark 192
White-browed Tit-Spinetail 114
White-capped Dipper 152
White-cheeked Pintail 36
White-collared Jay 146
White-collared Swift 80
White-crested Elaenia 120
White-eared Solitaire 154
White-fronted Ground-Tyrant 132
White-lined Tanager 164
White-rumped Hawk 50
White-rumped Sandpiper 64
White-rumped Swallow 148
White-tailed Shrike-Tyrant 134
White-tailed Tyrannulet 122

White-throated Hawk 52
White-throated Quail-Dove 70
White-throated Screech-Owl 74
White-throated Tyrannulet 122
White-tipped Dove 70
White-tipped Swift 80
White-tufted Grebe 40
White-tufted Sunbeam 88
White-winged Black-Tyrant 130
White-winged Cinclodes 112
White-winged Diuca-Finch 178
Wilson's Phalarope 64
Wood Stork 42
WOODPECKERS 98
Wren-like Rushbird 110
WRENS 150
Xenodacnis parina 174
Xenops rutilans 110
Xiphocolaptes promeropirhynchus 108
Yellow-bellied Elaenia 120
Yellow-bellied Seedeater 180
Yellow-billed Cacique 192
Yellow-billed Cuckoo 72
Yellow-billed Pintail 36
Yellow-billed Teal 36
Yellow-billed Tit-Tyrant 122
Yellow-browed Sparrow 182
Yellow-crowned Night-heron 42
Yellow-olive Tolmomyias (Flycatcher) 126
Yellow-scarfed Tanager 166
Yellow-throated Tanager 166
Yellow-whiskered (Short-billed) Chlorospingus 184
Yellow-winged Blackbird 192
Yungas Pgymy-Owl 76
Zenaida auriculata 70
Zentrygon frenata 70
Zimmerius bolivianus 124
Zonotrichia capensis 182